Walden Warming

Walden Warming

CLIMATE CHANGE
COMES TO THOREAU'S WOODS

Richard B. Primack

The University of Chicago Press ❧ *Chicago & London*

The University of Chicago Press, Chicago 60637
The University of Chicago Press, Ltd., London
© 2014 by The University of Chicago
All rights reserved. Published 2014.
Paperback edition 2015
Printed in the United States of America

23 22 21 20 19 18 17 16 15   2 3 4 5 6

ISBN-13: 978-0-226-68268-6 (cloth)
ISBN-13: 978-0-226-27229-0 (paper)
ISBN-13: 978-0-226-06221-1 (e-book)

DOI: 10.7208/chicago/9780226062211.001.0001

Library of Congress Cataloging-in-Publication Data

Primack, Richard B., 1950– author.
Walden warming: climate change comes to Thoreau's woods / Richard B.
Primack.
pages; cm
Includes bibliographical references and index.
ISBN 978-0-226-68268-6 (cloth: alk. paper)—ISBN 978-0-226-06221-1
(e-book) 1. Plants—Effect of global warming on—Massachusetts—
Walden Pond State Reservation. 2. Animals—Effect of global warming
on—Massachusetts—Walden Pond State Reservation. 3. Plants—Effect
of global warming on—Massachusetts—Concord. 4. Animals—Effect
of global warming on—Massachusetts—Concord. 5. Climatic changes—
Massachusetts—Walden Pond State Reservation. 6. Climatic changes—
Massachusetts—Concord. 7. Thoreau, Henry David, 1817–1862. I. Title.
QH105.M4P75 2014
577.2709744—dc23   2013038942

Frontispiece: Statue of Henry David Thoreau in front of a replica of his cabin on the edge of Walden Pond; photo by Richard B. Primack and Abraham J. Miller-Rushing.

♾ This paper meets the requirements of ANSI/NISO Z39.48-1992
(Permanence of Paper).

Contents

Preface

I wish so to live ever as to derive my satisfactions and
inspirations from the commonest events, every-day phenomena,
so that what my senses hourly perceive, my daily walk,
the conversation of my neighbors, may inspire me,
and I may dream of no heaven but that which lies about me.

THE JOURNAL OF HENRY DAVID THOREAU,
MARCH 11, 1856

EVERY SPRING FOR THE PAST ELEVEN YEARS, I have walked around Walden Pond and other places in Concord, Massachusetts, recording the first blooming dates for all of the wildflowers. I am repeating the same observations made a century and half ago by the author and philosopher Henry David Thoreau. On May 11, 1853, Thoreau recorded the first open flower of highbush blueberry. Its distinctive white tubular flowers are easy to observe. In subsequent years, he recorded the first blueberry flowers in Concord between May 14 and 19.

If Thoreau went looking for the first blueberry flowers of Concord in mid-May today, he would be too late — some bushes would be covered with flowers, while others would have only a few stragglers left hanging among the young green fruits. Since the 1850s, the first blueberry flowering has shifted three weeks earlier — they now generally open during the last two weeks of April. However, in 2012, after a record warm winter, blueberry bushes began to flower on April 1, six weeks earlier than in Thoreau's time. When a historical perspective is combined with modern observations, one thing becomes clear: climate change has come to Walden Pond.

By the late 1980s, I and many other biologists began to see that our world was being transformed and often severely degraded by human activities. I increasingly came to believe that botanists, wildlife ecologists, and other field naturalists should use their knowledge of the living world to help preserve those wild species of plants and animals that do so much to lend beauty and grace to our human existence, and that provide food, timber, medicine, and ecosystem services that are essential to human well-being. So I switched from researching the ecology of plants and animals in undisturbed natural areas to documenting human impacts on the diversity of the living world and finding ways to conserve and restore the species and ecosystems that remain. During the 1990s, this meant studying and writing about the effects of logging on the island of Borneo. But beginning in 2002, I made another second significant shift in my career. I decided to apply my knowledge of plants and natural history to an examination of the effect of warming temperatures and other aspects of climate change on plants and animals.

I soon learned that Thoreau and other naturalists had observed the plants and animals of Massachusetts, and these observations could be updated and reanalyzed to gain perspective on the impacts of a warming climate. I even came to regard Thoreau as a scientific colleague and considered adding his name as a coauthor to my research publications. This book describes my adventures with my students and colleagues as we studied the local effects of a warming climate, building upon the foundation established by Thoreau.

Richard B. Primack
Boston University, 2014

Ice-out at Walden Pond; photo by Tim Laman.

I frequently tramped eight or ten miles through the deepest snow to keep an appointment with a beech-tree, or a yellow birch, or an old acquaintance among the pines.

THOREAU, *WALDEN*

1. Borneo to Boston

IN 2001, I WAS CONDUCTING FIELDWORK IN Bako National Park, a rugged rain forest landscape of jagged rocks and cliffs on the northwest seacoast of Malaysian Borneo. I was studying how hundreds of different trees could coexist in a single stand in the rain forest and the ways that selective logging was affecting both the mix of species and the structure of the forest. Trees were my focus, but a spectacular variety of animals and herbaceous plants confronted me at every turn. Each morning as I set out from the field station, I could hear large proboscis monkeys calling and moving through the forest canopy above me; rattan vines cascaded down from the trees; and fallen flowers and leaves of great variety littered the forest floor. However, even though Bako National Park was a paradise for a nature-lover like myself, it was impossible for me to ignore the reality that all around the park, as well as up and down the whole island, the forest was vanishing. It was being cleared to harvest timber and to make way for palm-oil plantations.

The clearing of rain forests in Borneo and across the Old and New World tropics is having devastating consequences for biodiversity, but the destruction of tropical forests around the globe is hastening a different kind of change: climate change. Rain forests fix carbon — they are among the largest carbon reservoirs on the planet — and when people clear and burn them, tons of carbon dioxide are released into the atmosphere. People can understand how habitat destruction leads to the loss of orangutans, hummingbirds, and orchids. But what they don't understand as well is how clearing trees in Borneo and the Amazon can affect climate on the other side of the world. Increasingly, as I measured trees and patrolled study plots, it was climate change that was on my mind.

I had made a turn earlier, with many other field biologists, to conservation biology. This relatively new branch of ecology developed in the 1980s as field biologists recognized that they could not just passively observe the destruction of the species and ecosystems they were studying. Many of us decided that we should take an active role in protecting what we observed, valued, and loved. Our role models were people like Jane Goodall, who changed from carrying out long-term studies of chimpanzees in Africa to leading global efforts on behalf of primate conservation; Dan Janzen, who has devoted his life to creating and managing the vast Guanacaste National Park in Costa Rica; and Dian Fossey, killed at her field station while aggressively defending gorillas against poaching. These three pioneers were among the champions who inspired thousands of young people and established scientists to enter the new field of conservation biology and become involved in protecting the diversity of life on Earth.

Since earning my doctorate from Duke University in 1976, I'd been following the traditional career trajectory for a field botanist. I'd been studying forest ecology in Malaysia since 1980, and I even thought of it as a second home, since that is where my wife, Margaret, and her family are from. Starting in 1990, I had increasingly begun to focus my tropical research on conservation biology. By 2001, I found myself making a second turn, to the study of climate change. Being a botanist, I would focus my attention on looking for the effects of climate change on plants, while keeping my eyes open for the impact of climate change on birds and other animals. I loved working and living in Borneo, because of its astonishing natural history and wonderful people, but I concluded that as a scientist I needed to influence opinion on environmental issues in the United States. To build a compelling case for the effects of climate change, I would need evidence of the impacts of warming temperatures not in the far-off rain forests of Malaysia, but back in the United States. People might not know or care much about proboscis monkeys and pitcher plants on the other side of the world, but they would not be able to ignore the evidence from places closer to home and species they know and care about.

After all, the whole point of global warming is that it's a *global* phenomenon that affects every place on planet Earth. Indeed, showing that the effects could be felt in our own backyard, literally, was part of the point I wanted to make. It's easy for people to be nonchalant about or ignore a distant problem, but harder to remain unconcerned when it's up close and personal.

So, after twenty-one years of studying tropical rain forests, I decided not to return to Malaysia for another field season. I would no longer make the long flight across the Pacific, walking through the national parks of Borneo, looking up at the giant trees with awe. Instead, I would become a detective, looking for signs of climate change in the field books of long-dead naturalists and among the observations of a new generation of ecologists. All of my field notes, data sheets, maps, and computer files from Borneo were put into a file cabinet in my lab, where they will remain until either I decide to resume my tropical work or some colleague or graduate student decides to take over the project.

Late in the winter of 2001–2, I returned to Boston and began my search for signs of climate change. I didn't know what I would find. Along with the chilly weather, I was encountering some cold receptions from my colleagues. Some of them wondered if it was really such a good idea for me to be changing my research direction at this point in my career. They pointed out that I was fifty-two years old and known as a tropical biologist; wouldn't it be better for me just to keep working in my narrow specialty? Other colleagues were skeptical that I would actually find anything of value in old field records and regarded my quest as doomed from the start. And most alarmingly, when I began to talk about my new interests, a number of researchers told me that the National Science Foundation and other government agencies would never fund this kind of wild-goose chase and that I had better find other projects if wanted to secure grant support. After a year of searching, I still had not made any research progress or figured out how to fund this new line of work.

Just when I was starting to have major doubts about my decision to change directions, into my office in early 2003 came Abe Miller-Rushing, a prospective graduate student looking for an innovative project for his doctoral dissertation. I was able to convey my excitement to Abe, and he soon agreed to work with me to search for evidence of climate change. I didn't tell Abe at the time, but it raised my level of confidence that at least one other person thought this idea might actually work.

We Find Thoreau's Records

The plan I'd outlined back in Borneo, and now conveyed to Abe and my Boston colleagues, was to look closely at ecosystems in Massachusetts to see if I could observe the evidence of climate change here, right under our noses. Now that I was back in the United States, I began by limiting my focus to eastern Massachusetts, where I owned a house and had worked since 1978 as a professor at Boston University.

As the accumulated snow of 2003 began to melt and the days started to lengthen, I felt that the clock was ticking. I needed to plan my research before the flowers bloomed and the trees leafed out. The best idea that I could come up with was to record the flowering times of plants in Concord and at the Arnold Arboretum in Boston. This would generate some useful information and keep my undergraduate research assistants busy for the summer. At the same time, I would begin looking for evidence that plants, birds, insects, and other animals were changing in response to a warming climate. Plants and animals become active in the spring and dormant in the autumn; my aim was to discover whether these dates were changing and whether the change was apparent over the span of decades. If so, how could I find out? To shed light on this question, we needed data. Abe and I began searching for any old records we could compare to modern inventories.

In the spring and summer of 2003, we began to have some luck. We located journals documenting when migratory birds arrived in the spring over the past forty years. But finding records further back than 1970 proved difficult, and we still had not found any old records of plant flowering times. The whole research enterprise depended upon having a data set from fifty or more years ago to which we could compare modern numbers. But did the older data sets even exist? By the end of our first field season, we had lots of information on modern flowering times. However, despite months of scouring every library and archive we could think of, as well as polling colleagues and writing notices for newsletters and websites, we still hadn't found the old data sets that we needed. Then, our luck changed.

Our "eureka moment" came a year into our study, when in the fall of 2003 Phil Cafaro, a friend of mine who is a professor of environmental philosophy, asked me if I knew about the flowering-time observations that Henry David Thoreau had made in Concord during the mid-1800s. As I listened with amazement, I realized that this was exactly the information we had been seeking. Phil told me that Thoreau had made observations in Concord of plant flowering times, as background to a book he was planning but never wrote on the change of seasons. It turns out that this list of flowering dates is well known to Thoreau scholars, and I was able to get a copy within days from Brad Dean, an independent Thoreau scholar living in New Hampshire. The list was everything I could have wished for: the first flowering dates of plants in Concord for the years 1851 through 1858. Thoreau, in addition to being one of the key figures of transcendentalism and a pioneer of the American conservation movement, had also made annual quantitative observations of an entire plant community. We now had our baseline data to see how plants had changed their flowering dates over time.

"A Self-Appointed Inspector"

During the fall, as I talked with Brad and then later with Jeff Cramer of the Thoreau Institute and other Thoreau scholars, I learned that Thoreau had kept extensive journals in which he recorded the seasonal phenomena he observed during his walks along the forested shores of Walden and surrounding areas of Concord. He noted the dates of flowering and leaf-out times, when birds arrived in the spring, the date of ice-out at Walden Pond, and other natural phenomena in Concord. This came as a complete surprise; I'd expected we might find records in the archives of a nineteenth-century natural history society, but not among the writings of one of the most revered and original writers in the American literary canon.

I'd read *Walden* in college, and I knew that Thoreau was considered one of the founding fathers of the modern conservation movement — but I had never known (nor had any other biologist that I talked to) that he'd made detailed observations of local flora and fauna while he lived out by Walden Pond between July 1845 and September 1847 and for eleven years thereafter. But these weren't just any written observations.

Thoreau had prepared tables precisely documenting the annual flowering times of over three hundred plant species. Species occupied row after row on successive pages of surveyor's paper, with years listed in the column heads at the top of each sheet. Thoreau had created these tables by extracting data from all of his journals. We had found our dream data set, and other data sets were to follow.

Thoreau's natural history tables had never been published, with their existence known to only the small circle of Thoreau scholars. Once we found out about them, my colleagues and I realized that these records offered us a perfect window into the local ecology of eastern Massachusetts a century and a half ago.

In the days after we learned of the Thoreau data, I met with Abe Miller-Rushing to plan our next steps. We were excited and cautiously hopeful. There were unknowns, as there always are when setting out on a new research project. Could we find a way to use Thoreau's observations to show whether the effects of global warming could be seen in Concord and eastern Massachusetts? At a more basic level, we had no idea whether such a project had ever been attempted and only a vague idea of how to go about it. It seemed impossible that no ecologist or botanist had ever seen or worked with these records before. And even if we *could* document changes in such phenomena as the appearance of spring flowers, and even if we *could* attribute them to climate change, what should we do about it?

The most obvious path was to build up our own data bank of observations of the same phenomena, so that we could compare the two sets of data and look for similarities and differences. We knew that the landscape of Concord had changed greatly in the 160 years between Thoreau's observations and ours, but we also recognized that Concord, more than most suburban areas, was extremely well-protected by government agencies and private land trusts. We would have the opportunity to study the same species that Thoreau had observed in some of the same woodlands, pond edges, and river meadows, following in his footsteps.

> For many years I was self-appointed inspector of snow storms and rain storms, and did my duty faithfully; surveyor, if not of highways, then of forest paths and all across-lot routes, keeping them open, and ravines bridged and passable at all seasons, where the public heel had testified to their utility. (Walden, 17)

And so in 2003, my students and I became the self-appointed inspectors of climate change in eastern Massachusetts, rediscovering old, forgotten, or underappreciated data sets, interviewing bird-watchers, and recording flowering dates of different plants. For these observations, we hiked the landscape of Concord two to three days each week every spring, noting the earliest dates that particular flowers opened up (bud-burst), just as Thoreau had done, as a gauge of the impacts of climate change.

At the same time that we were beginning our own observations, I was delving into Thoreau's writings, to see what else he had to say about weather phenomena in the mid-nineteenth century. In the 1840s, the Industrial Revolution was just beginning to take hold in America. The boom in the use of coal, and its associated production of carbon dioxide, wouldn't come until the 1870s. Thus when Thoreau began his experiment at Walden, he was more often than not writing about episodes of extreme cold, frosts, and thick ice, rather than our present focus on warming trends. His age had different preoccupations, and people were more concerned with freezing to death than suffering from extreme heat.

On Thin Ice

In the 1840s, when Thoreau was living at Walden Pond, the idea of global warming did not yet exist, nor was there any reason for anyone to propose it. The farmers and townsfolk of nearby Concord, Massachusetts, were far more concerned with cold conditions. An early spring was a phenomenon people would have greeted with pleasure, as it meant relief from heavy snows and bone-chilling temperatures.

The mid-nineteenth century was a particularly cold period, deeply felt by a largely rural society. Frosts during the summer could kill the crops upon which farmers depended for food and income. A late frost on the night of June 12, 1846, killed the beans, tomatoes, squash, corn, and potatoes that Thoreau had planted in his field. Severe winter cold could also kill livestock in unheated barns as well as wild game, such as deer, that supplied supplemental meat. During an unusually cold winter, families might run out of firewood trying to heat their houses and would experience great hardship or freeze to death. In his journal, Thoreau wrote about the extraordinary cold weather of 1810, when he was three years old:

*Mother remembers the Cold Friday very well. . . . The people in the kitchen . . .
drew up close to the fire, but the dishes which the Hardy girl was washing froze
as fast as she washed them, close to the fire.* (*Journal*, January 11, 1857)

Had Thoreau been able to look into the future, I imagine he would
have been dismayed by some of the changes that have occurred in Concord. In particular, the effects of climate change and rising temperature
are now affecting his beloved Walden Pond. In New England, one place
where global warming can be seen clearly is its lakes and ponds, now covered in winter with no more than a thin veneer of ice where before there
was a thick ice covering.

The failure of thick ice to form on bodies of water is a phenomenon
that has happened more frequently in recent decades. In the 1950s and
early 1960s, when I was growing up in Newton, Massachusetts — a suburb of Boston not far from Concord — ice-skating was a popular pastime,
with hundreds and even thousands of people gathering to ice-skate on
Crystal Lake and Hammond Pond. I remember crowds of people gliding
across the endless glassy-smooth lake surfaces, and then gathering around
log fires to warm their frozen hands. A section of the lake was set aside for
pickup games of hockey. In those years, winter temperatures were typically in the teens or single digits for weeks at a time, and the ice was very
thick.

During the late 1980s and early 1990s, when I wanted to bring my
own children ice-skating in such a natural setting, it was hard to find a
suitable pond or lake in Newton. The winters were now too mild; without the long cold snaps of my boyhood, ice of sufficient thickness no
longer formed. The thin ice brought risks of people falling through and
drowning. The Newton city government moved to restrict skating at the
traditional lakes and ponds, and urged people to use public skating rinks.
Many citizens, however, still longed for the natural experience of skating on a lake or pond. The city of Newton responded by designating two
spots for ice-skating, Bullough's Pond and Ware's Cove on the Charles
River. We enjoyed ice-skating trips to these places when our children were
small. But after one or two seasons, the city had to abandon this plan.
Too many winter days were above freezing, and the ice was just not thick
enough to support safe skating.

In the autumn of 2011, a new fad hit the Boston area. Park department workers and parent volunteers built large wooden frames, 10 to 30 feet across, on open ground in town centers and playgrounds. While the frames resembled giant sandboxes, the actual plan was to flood the low frames with water during the freezing days of winter, creating a smooth ice-skating surface for children. However during the mild winter of 2011–12, daytime temperatures never consistently went below freezing. The skating rinks were never flooded and went unused.

Ice at Walden Pond

Ice too thin for skating and walking would have shocked Thoreau — it just wasn't within the realm of his personal experience. On February 23, 1854, Thoreau's journal stated that he walked to the middle of Walden Pond and measured the thickness of the ice as 17 inches. And on March 11, 1856, he walked out to the middle of Walden Pond and cut a hole in the ice. He measured a top layer of 3 to 4 inches of crusted snow, 11½ inches of snow ice, and 12¾ inches of solid ice. He concluded, "Snow and ice together make a curtain of twenty-eight inches thick now drawn over the pond. Such is the prospect of the fishes!"

Contrast Thoreau's experience with this recent past year. On February 20, 2012, I took my son Jasper for a walk around Walden Pond. The day was sunny, and the temperature was in the mid-40s, which was typical for the previous week. As we left the parking lot and headed to the pond, we were greeted by an astonishing sight. Walden Pond was not frozen. It was an open, glittering surface of water, with some thin ice lingering in a few of the shaded coves. The foot-thick sheet of ice that Thoreau had observed 156 years earlier was a phenomenon of the past. Just like the melting of ice sheets and glaciers in polar regions, we are now experiencing the disappearance of ice on a smaller scale at Walden Pond.

A keen observer of ice, Thoreau knew exactly when to trust it or avoid it. On February 4, 1857, he wrote:

> . . . as I stood on Walden, drinking at a puddle on the ice, which was probably
> two feet thick, and thinking how lucky I was that I had not got to cut through
> all that thickness, I was amused to see an Irish laborer on the railroad, who

had come down to drink, timidly tiptoeing toward me in his cowhide boots, lifting them nearly two feet at each step and fairly trembling with fear, as if the ice were already bending beneath his ponderous body, and he were about to be engulfed. "Why, my man," I called out to him, "this ice will bear a loaded train, half a dozen locomotives side by side, a whole herd of oxen," suggesting whatever would be a weighty argument with him. And so at last he fairly straightened up and quenched his thirst.

In addition to measuring the thickness of the ice, Thoreau regularly noted the day of ice-out at Walden Pond — that is, the day that the ice was no longer present over most of the pond surface — every year for fifteen years between 1846 and 1860. Such a careful set of observations could allow us to detect the fingerprint of a warming climate. According to Thoreau's records, ice-out at Walden Pond occurred as early as March 15 and as late as April 18, a five-week range, with an average ice-out date of April 1.

In recent years, park rangers and volunteers at Walden Pond State Reservation have continued this tradition of observing ice-out dates at the pond, with the results posted online. During the fifteen years between 1995 and 2009, ice-out ranged from February 22 to April 12, a range of seven weeks, with an average of March 17. The average date over the past fifteen years is two weeks earlier than in Thoreau's time. If Thoreau had been around during the past twenty years, using his nineteenth-century records, he would have been able to detect the effects of climate change in the shifting ice-out dates at Walden Pond. And what is even more remarkable, the first three months of 2012 were the warmest first quarter of the year ever recorded in Boston, with temperatures about 9 degrees above normal. During this exceptional winter, the temperatures were so mild that Walden remained free of ice for most of the winter. Even when the pond had a thin skin of ice in the early morning after a cold night, the ice soon melted and the watery surface reappeared. A stable ice surface only formed on January 19, and ice-out for the season was officially recorded only ten days later on January 29, six weeks earlier than the ice-out in Thoreau's time and three weeks earlier than any recent year. These mild winter days and the absence of an ice surface would have astonished Thoreau, who was accustomed to harsh winters and thick pond ice.

These simple observations from Walden Pond made by Thoreau, the park ranger, and volunteers without any fancy equipment, data loggers, or high-tech sensors on satellites show that local events in Concord are already responding to the regional changes in weather that are occurring in eastern Massachusetts. Not only is the weather getting warmer, but it is also affecting bodies of water like Walden. In turn, the timing of ice-out affects when waterfowl, such as ducks and loons, are able to land on the surface of the pond and begin foraging for food. It affects when aquatic insects such as midges, mosquitoes, and dragonflies — which have overwintered underwater as nymphs or in other juvenile states — are able to complete their metamorphosis, emerge from the pond, and begin to fly. It determines when fish are able to start coming to the pond surface to catch these insects. And the thick black northern water snakes stay in hibernation inside rock crevices until the ice melts and the water warms enough for them to wake up and begin hunting. So ice-out is a physical phenomenon that has wide-ranging effects on the pond's fauna and its food chain.

But does air temperature really determine when ice-out occurs at Walden? Can it explain the earlier ice-out that we see today? The results of our analysis provide a clear answer to this question. The ice-out at Walden is highly correlated with temperatures in January and February, the two months preceding ice-out in most years. Ice-out is earlier than usual in years where January and February temperatures are warmer than average, and it occurs later in years with below-average January and February temperatures. We can even predict how early (or late) ice-out will be based on temperatures in those months: ice-out occurs three days earlier for each single degree Fahrenheit increase in temperature during the first two months of the year. In Thoreau's time, the temperatures in the Boston area averaged 25 degrees Fahrenheit in January and February, whereas Boston's present temperature averages 28 degrees. This temperature increase of 3 degrees for these two months largely explains the earlier ice-out that has been observed at Walden from Thoreau's observations to the present. These results also demonstrate the exquisite sensitivity of ponds to changes of temperature of just a few degrees.

We also know that other bodies of water, such as lakes and streams, and ice formations, such as glaciers, share this sensitivity to temperature variation. Changes of around 8 degrees in late winter temperatures can determine whether Walden Pond will thaw in early March or early April, with enormous implications for the timing of all of the living creatures that live in the pond or in some way depend on it. With winter temperatures in 2012 an astonishing 9 degrees warmer than average, Walden Pond was completely free of ice for most of the winter, something that certainly would have surprised Thoreau.

Thoreau did not use his measurements of ice-out dates to look for long-term changes, because such changes occur over many decades or several lifetimes. He certainly was not aware of the rising carbon dioxide concentrations in the atmosphere as the United States entered the Industrial Revolution, already well under way in England, nor did he discuss this situation in his journals.

Thoreau's observations of ice-out were part of a long tradition carried out in many places throughout the temperate world. Ice-out dates were highly significant to commerce: ship captains could not sail into frozen harbors or across frozen lakes, while overland hauling of heavy cargoes, such as timber and stones, was done more easily in the winter by sledges drawn by draft horses or oxen across the ice-covered waterways (especially if the alternative was to haul across water by boat or going extra distances to bridges or fords in the summer). For such reasons, naturalists, meteorologists, and sailors often recorded the time that bodies of water froze in the early winter and thawed again in the late winter or spring. The ice itself was a valuable commodity, with ice from New England's ponds and lakes being harvested in winter and shipped all over the world primarily for keeping food and beverages cool in restaurants, food markets, dairies, and breweries.

We now recognize that changes in winter and early spring temperatures can be seen in the changing dates of ice-out, or thawing, at Walden Pond and other bodies of water around the world. The timing of thawing in the spring of a large body of water can be viewed as an average of the temperatures for the past few weeks or months. Walden Pond, which is over one hundred feet deep in the middle, does not respond to one or a few days of warm weather. It needs a prolonged period of warm weather to melt the ice on the surface and reveal the water below. A delay of spring warmth will similarly delay the melting of the ice. Thoreau was aware of the value of Walden as an indicator of the arrival of spring weather, writing:

This pond never breaks up so soon as the others in the neighborhood, on account both of its greater depth and its having no stream passing through it to melt or wear away the ice. It indicates better than any water hereabouts the absolute progress of the season, being least affected by transient changes in temperature. (*Walden,* 324)

Thoreau would certainly have been satisfied to find that his simple observations of ice-out at Walden Pond could be used to show so clearly the effects of a warming climate on the environment.

High water in Walden Pond surrounds a tree after spring rain; photo by S. B. Walker.

And then the rain comes thicker and faster than before,
thawing the remaining frost in the ground, detaining the migrating bird;
and you turn your back to it, full of serene, contented thought,
soothed by the steady dropping on the withered leaves, more at home
for being abroad, more comfortable for being wet.

THOREAU, *JOURNAL*, JANUARY 27, 1858

2. A Hard Rain

OUR RESEARCH IN AND AROUND Walden Pond was taking place during an intense political and cultural debate about the reality of climate change. Between 2003, when our research began, and the present, the American public and the scientific community were being presented with unambiguous evidence of how climate change was affecting global temperature, with resulting impacts on glaciers, polar ice caps, and even the vast oceans. These changes in temperature and environments were affecting the abundance and distribution of animals and plants species around the world. As a result of this evidence, the scientific community reached a broad consensus that climate change was happening; that it was being caused by human activities, in particular the burning of fossils fuels and cutting down of tropical forests; and that sweeping governmental action was needed to halt the production of greenhouse gases in order to slow down global warming.

These findings were conclusively documented and presented in the 2007 report of the Intergovernmental Panel on Climate Change, an international committee of hundreds of the world's leading scientists. And yet at the same time, climate change became part of the political debate in the United States rather than being addressed by environmental policy. Small numbers of scientists and lobbyists, often linked to conservative political groups and think tanks funded by the energy and transportation industries, confused the issue by arguing that the evidence for climate change was flawed and inconclusive. These climate change deniers were successful in creating doubt in the public and political spheres, with the result that no clear and compelling mandate has emerged in the United States or at the international level to address the causes of climate change. The two international climate change meetings in Copenhagen in 2009 and Rio de Janeiro in 2012 did not lead to any enforceable action plan to reduce the production of greenhouse gases.

Even though people may refuse to take notice of it, climate change will not be easy to ignore. Rather than the gradual warming trends that are the real signature of climate change, it is often dramatic weather events that bring home the reality of shifting patterns. In Paris in 2003, temperatures reaching 104 degrees Fahrenheit led to the deaths of over ten thousand people, mainly senior citizens. This record heat wave had the French talking about global warming's effects and convinced them of its reality. For residents of New Orleans, the devastation wrought by Hurricane Katrina in 2005 raised awareness of how climate change could directly impact their lives. Russians experienced uncontrolled wildfires spreading across a drought-ravaged landscape, destroying dried-out crops and causing alarm in the summer of 2010. American farmers remember the summer of 2012 for the extreme drought caused by record high temperatures and little rain. And New York and New Jersey residents are still recovering from the damage associated with Hurricane Sandy, including the flooding of lower Manhattan and many of the subway tunnels. But for many citizens of greater Boston, the most striking weather event since the Great Blizzard of 1978 was the massive rainstorm that fell on the region over three days starting on March 13, 2010.

I had visited Walden Pond the day before, and the pond was still largely frozen. There was a band of open water along the northeastern edge of the pond where the low winter sun could warm up the sandy beach without interference from overhanging trees. Based on my past experience, I guessed that the pond still had a week or two to go before ice-out. This would put the ice-out for 2010 somewhat later than March 17, which was the average for recent years, and within the range of Thoreau's observations. It looked to me that this might be a year that did not show the effects of climate change.

The next day, quite unexpectedly, a tremendous rainstorm hit Boston, depositing ten inches of rain over three days, making it the heaviest concentrated rainfall to hit the city in more than a century. All at once, the deluge melted the accumulated snow and ice throughout the city, adding to the rainfall all the water that had been held in its frozen state. After the storm ended, I returned to visit Walden Pond on March 16 to find the surface rippling with water and no ice to be seen. It had all melted during the storm. Thoreau had also observed such sudden disappearance of ice on Walden Pond, noting one day in March:

I looked out the window and lo! where yesterday was cold gray ice there lay the transparent pond already calm and full of hope as on a summer evening, reflecting a summer evening sky in its bosom, though none was visible overhead, as if it had intelligence with some remote horizon. (*Walden*, 338)

This March rainstorm had disproved my prediction for a late ice-out for 2010, and changed the ice-out date to one fitting the pattern of recent decades, and several weeks earlier than ice-out in Thoreau's time. The storm had dramatically changed the appearance of the pond: runoff had raised the water level all along the shore, submerging the path in many places, and one of the most popular sandy beaches lay under several feet of water. The water level remained exceptionally high throughout the summer, even though the remaining summer months saw lower than usual rainfall.

Thoreau had also noted such changes in the level of Walden Pond, with cycles of years of high water followed by years of low water.

The Pond rises and falls, but whether regularly or not, and within what period, nobody knows, though, as usual, many pretend to know. It is commonly higher in the winter and lower in the summer, though not corresponding to the general wet and dryness. (*Walden*, 197)

Such alternations of high-water years with low-water years are important to the ecology and appearance of the pond. High water strangles encroaching trees and shrubs, helping to keep the open appearance of the pond. On December 5, 1852, Thoreau wrote in his journal:

This great rise of the pond after an interval of many years, and the water standing at a great height for a year or more, kills the shrubs and trees about its edge, — pitch pines, birches, alders, aspens, etc., — and falling again, leaves an unobstructed shore. . . . This fluctuation, though it makes it difficult to walk around when the water is highest, by killing the trees, makes it easier and more agreeable when the water is low.

Thoreau's observation from 160 years ago could easily have been provided by a consulting hydrologist paid by the state of Massachusetts to determine what to do about the abnormally high water levels in 2010 — which is to say, do nothing, because the action of the water will actually benefit the pond by pushing back the encroaching brush. And that is exactly what has happened. Just as in Thoreau's time, the high water of 2010 killed most of the pine trees at the water line along with many of the alders and other woody plants that had their roots covered by water in March, April, and May 2010.

Nor'easters in New England

The rainstorm of March 2010 illustrates what the changing climate will likely deliver to eastern Massachusetts with increasing frequency in coming decades. This storm was a classic "nor'easter," a distinctive weather pattern of coastal New England that occurs between autumn and early spring. A nor'easter often originates as a low-pressure area in the warm waters of the Gulf of Mexico, then re-forms off the coast of the eastern United States. As it moves north, this type of storm gains strength from the difference in temperature between the warm storm mass and the colder North American continental temperatures. The air in a nor'easter rotates counterclockwise, lashing the coast with powerful wind-driven rains from the northeast as the storm passes through.

Nor'easters typically last anywhere from a few hours to a day at any one location, usually bringing a short burst of intense snowfall or heavy rain. Sometimes nor'easters can last a few days, with a large accumulation of snow and rain when a high-pressure system in eastern Canada slows the northward progress of the storm. Growing up in Boston, I had gotten used to the idea of a heavy nor'easter rainstorm as being about an inch or two. Of course, intense summer thundershowers might drop that much in half an hour, but that was a different type of storm. A nor'easter arriving in the winter can sometimes become a blizzard, with strong winds depositing up to a foot or more of snow.

These nor'easters are well known to New Englanders, and while well respected, they generally are not feared. Yet at the same time, climatologists are predicting that all such storms, including hurricanes and thunderstorms, will become more powerful in the future, bringing higher winds and heavier rainfall. Warming temperatures bring more energy into storm systems, leading to storms both more frequent and more intense than we've been accustomed to seeing. Katrina and other hurricanes of the past decade have received attention as the first noticeable examples of such changes in the weather (that is, the short-term variations in meteorological phenomena, as opposed to the long-term trends that we call climate). New England has experienced some massive nor'easters in the past, and such "once in a generation" events are predicted to become more common in coming decades.

In Thoreau's time, people in New England still remembered the Great Snow of 1717 — a storm that took place a full century before Thoreau's birth — when five feet of snow fell throughout the region over eight days, isolating people in their homes and killing livestock left outside. In many places the snow lay more than ten feet deep.

A more recent nor'easter, the Blizzard of 1978, buried most of the East Coast of the United States between Maine and Washington, D.C., with snow. In Boston the three-day storm dropped twenty-seven inches of snow, which drifted in places to fifteen feet and paralyzed the city for over a week.

The March 2010 nor'easter brought rain. By the time it ended, many town centers across New England would be underwater, entire neighborhoods would be flooded, major highways would be closed, rivers would be overflowing their banks, and 450,000 people would lose their electrical power.

The morning of March 13, 2010, dawned with heavy clouds, and it rained steadily throughout the day. The rain continued that evening and through the night, sometimes light, sometimes heavy, but always steadily. Instead of moving on past Boston, the nor'easter had stalled offshore, hammering away at the city.

Sunday morning, March 14, when my family looked out our back windows, we confronted an alarming sight. Not only was it still dark and raining at 8 a.m., but our backyard was filled with water and now resembled a small pond. Our vegetable garden, which occupies the back half of the yard, was underwater, with only the tops of last year's tomato stakes breaking the surface. Our backyard adjoined the backyards of the neighbors to our right and left and the backyard of a house behind ours on the next street over, to form a basin some fifty by thirty yards. Eighty years ago, before the houses were built, this depression had been a swamp.

Along the fence line, you could still see remnants of that swamp vegetation — red maples, jack-in-the-pulpits, and skunk cabbages — a reminder of its ecological history.

During past nor'easters and annual spring snow melts, this low, oval basin would fill with water in the back half of the yards. The water would then gradually recede over the course of a few days. I had always assumed that the water drained off via a stream channel on the neighboring street, but I had never seen the stream myself.

On this Sunday morning, however, the water in the basin was higher than it had been in my fifty-five years of living in the house. The water filled our entire backyard and was creeping up to within ten feet of our basement door. Already the water level was several inches above the level of the basement floor of my right-hand neighbor's house and was close to the basement level of the left-hand neighbor's house. And across the fence, the water level was two to three feet higher than the basement of that house. This meant that two families' houses were already flooded, and our house and the house to the left were about to be!

Part of me was sure the rain couldn't last. But the rain kept falling, and hour by hour the floodwaters kept rising toward our house. Finally, at 5 p.m., I telephoned our neighbors and suggested that we meet outside to determine why we were having such massive flooding and to figure out what could be done about it.

Fifteen minutes later, we met in front of the flooded right-side neighbor's house. As we stood there in the cold, pouring rain, we realized that we had to answer two questions: First, why was the basin filling up so quickly with water, and, second, why wasn't the water leaving the basin, perhaps via a streambed on the neighboring street? And from a practical point of view, how could we prevent water from flowing into the basin, both now and in the future, and how could we ensure the water drained more quickly? Of course, during this meeting and subsequent discussions, we never asked the larger question of why we were having such an enormous and unusual rainstorm in the first place. The topic of global warming never came up — we were, after all, concerned with the more immediate problem of saving our homes.

Looking at the expanse of water, we recognized that we were in the midst of a small-scale natural disaster that mirrored the catastrophic flooding that was overwhelming all of eastern Massachusetts. Our flooding, and all of the flooding of the region, was caused by a record-breaking rainstorm at the worst time of year. The rain was falling at a time when the winter snow and ice had just melted on the surface, and the ground was fully saturated with water. And when we walked over to the neighboring street, we observed that there was no streambed or outlet draining water from the basin. The water was being held in place by high ground that gently rose above the water's surface, and water could only drain out of the basin by seeping into the ground. However, during this late winter storm, the ground eight to twelve inches below the surface probably remained frozen, leaving the floodwater with nowhere to go. It was being held in place as if the basin were an impermeable ceramic bowl.

I also recognized that each of the four home owners adjoining the basin had contributed to the flooding that we were now experiencing. One family had built an extension to their home, filling in a section of the basin and decreasing its capacity to hold water. Another neighbor had added in many cubic yards of soil to their backyard to create a dry, raised area for a playground set for their children. The third family had recently paved and expanded the driveway next to their home, making the ground less permeable in the process. And our family had developed an elaborate vegetable garden in our backyard by building raised beds about two feet above the ground using a mixture of soil and leaves. I contemplated this silently as I stood there in the pelting rain with my neighbors. It was sobering to realize that each of us had diminished the water storage capacity of the basin, in some way making the flooding worse. How much we contributed to the problem is uncertain. Perhaps the two houses would still have flooded from the storm's high water even without our alterations to the basin. Then again, if we'd made no changes, perhaps the two houses would now be dry. With these uncomfortable realizations, I returned home, soaked and cold.

The rain continued through Monday, though with declining intensity, and gradually petered out by the late afternoon. The water level stabilized within inches of our basement. We were lucky. Our house had survived. But two of the families had not been so lucky and had extensive damage to their homes. Over the next two weeks, the water level slowly receded, remaining only in the low places near the back fences.

In all, ten inches of rain fell during the course of the three-day March nor'easter in 2010. A one-inch rainstorm is a heavy downpour, and this was the equivalent of ten such storms, back-to-back, without ceasing, over three days. This was one of the heaviest concentrated rainstorms in Boston's recorded history, and its effects were made much more severe by the already waterlogged, frozen ground. The monthly total, which included additional rainstorms and floods the following week, was an astounding fifteen inches of rain — about four times Boston's average for the month of March.

What could be done to deal with the problem of flooding? We were able to seal our basement door with sheets of plastic to keep out the leading edge of the rising water. But such superficial solutions would not have helped our two neighbors, who suffered severe damage to their basements. Because flooding might happen again if such storms become more common in a warming world, we might consider tearing down our houses, restoring the basin back to its original condition as a swamp, and then building new houses in an upland area not subject to flooding. Alternatively, if countries of the world agreed to reduce their greenhouse gas emissions, we might be able to return to a climate of less intense rainstorms.

But both of these solutions are impractical to deal with the immediate problem — tearing down existing houses in flood-prone places is generally unrealistic as it would be too expensive on a local scale (and it would likely encourage settlement on hillsides in a manner that would create other problems, such as deforestation, soil erosion, and landslides). The second solution, reducing the amount of greenhouse gases, is a monumental task on a global scale, one that will take decades for society to implement. While reducing the production of greenhouse gases is the only true long-term solution to the problem of climate change, it would not address the immediate problem: the risk of our neighborhood flooding again during storms coming in the next few years.

The most effective solution for our small neighborhood would be to dig an outlet for the basin through its western edge, so that excess water could drain away into an existing trench. This solution is not an option for us because the city would not allow such a change in the hydrology of a wetland, and the dozen families with houses backing onto the trench would vigorously object to the additional water that would put their properties at risk if the trench overflowed. This is the new reality of climate change for our region: higher temperatures, more intense rainstorms, the increased risk and reality of severe flooding — and short-term solutions that, if not planned carefully, may solve one person's (or community's) problem at the expense of another's well-being, causing conflicts.

What Would Thoreau Have Done?

Faced with our situation, what would Thoreau himself have suggested? It's hard to say, but I feel certain that he would *not* have suggested that we tear down our houses and move to higher ground. Thoreau understood that people need a place to live (although the house itself might be as much a burden as an asset, as he pithily observed in *Walden*). Most likely, his suggestion would have been pragmatic: alter your house and its surroundings as best you can to alleviate the problem, and learn to live with the part you cannot change. My house has "got" me in this respect — by choosing to live there, I also choose to live with the disadvantages that come with living next to a wetland that fills with water after heavy episodes of rain.

For all that he wrote about nature, Thoreau was not (as some have made him out to be) a radical, antisocial individual who shunned humanity and saw value only in wild places. Walden Pond itself was not a wilderness by any stretch of the imagination, nor as pristine as many now assume. In the 1840s, the region immediately surrounding Walden was a mixture of woodlots, secondary forests, pastures, and farmland, some of which had been abandoned and some of which were still actively in use. The forests around the pond had been heavily cut for firewood and timber and otherwise damaged, both by local people and by workers on the nearby Fitchburg rail line (in *Walden*, Thoreau refers to "the shanties which everywhere border our railroads," at least some of which were located fairly close to his own shack at Walden Pond). And ice was extracted from the pond itself, both before and after Thoreau lived there. In short, Thoreau was no hermit, searching out the distant wilderness to avoid being sullied by society's demands. Nor did he seek to be.

Thoreau's point in going to Walden was not so much to avoid being around people — and indeed, if you read his journals, you find that for all his solitary walks, he did still interact regularly with the local residents — but to find solitude when he wanted it, the company of the natural world when he wanted it, and still maintain a connection to society *because* he wanted it. *Walden* is not, in the end, about the wonders of nature or exploring the spirit of the natural world — it is about exploring the human spirit and its place within the natural world and human society.

Thoreau coined a wonderful new word in *Walden*: "realometer," a portmanteau word combining "reality" and "thermometer." Thoreau's realometer is an instrument that would allow someone to measure the reality or truth of our perceptions and ways of life. He imagined a realometer would allow an inquiring person to push past the "mud and slush of opinion, and prejudice and tradition, and delusion, and appearance . . . to a hard bottom" of fact (*Walden*, 96). Today in the United States and around the world, we need such a realometer to examine the truth of global warming and the associated effects of climate change. Despite the clear evidence that the world is already warming due to the burning of fossil fuels and clearing of tropical forests, most Americans still do not regard climate change as a priority. They are more concerned with economic issues that they think affect them more immediately and believe that the effects of climate change will only be felt in the future and at some faraway place and not where they live today. It turns out that Thoreau himself has provided us with a realometer in the form of his extensive journals of a century and a half ago. We can use Thoreau's own observations to measure how a warming climate has affected the animals and plants of Walden Pond and surrounding Concord.

When it comes to the problem of my flooded backyard, and the bigger problem of the global climate crisis that led to the flooding in the first place, Thoreau's perspective on human material and spiritual needs can be instructive. Our understanding of humankind's place within the natural world is skewed to such an extent that it threatens our ability to live; therefore, we need to seek a new understanding. And it is fitting that one way of reaching that understanding of the impact of climate change on the environment is by using Thoreau's own observations as a realometer to test for the truth of climate change.

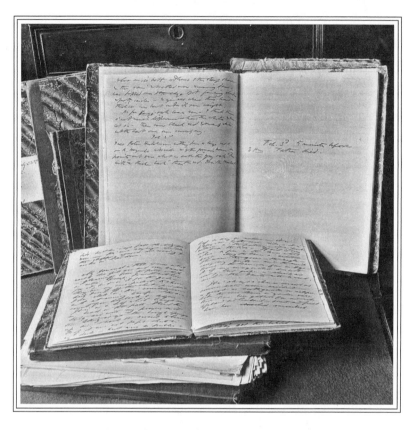

Thoreau's journals; Harold Wendell Gleason photo 1901.19 taken
February 26, 1901, at the home of E. H. Russell, Worcester, Concord,
Mass., Special Collections, Concord Free Public Library.

A lake is the landscape's most beautiful and expressive feature. It is earth's eye; looking into which the beholder measures the depth of his own nature.

THOREAU, *WALDEN*

3. Thoreau, Scientist

LOOKING OUT AT THE PLACID, DARK BLUE WATERS of Walden Pond on a late summer afternoon, the problems of the world seem far away. Except for a small sandy beach and a bathhouse at the eastern edge, the entire perimeter is thickly forested, with few signs of human influence. From this viewing place, it is easy to imagine that nothing has changed at Walden Pond in hundreds, if not thousands, of years. Perhaps the Native Americans hunting deer on these shores, the Revolutionary War minutemen rushing forward to do battle with the British redcoats, or the local soldiers training for the Civil War saw the same scenery at Walden Pond.

Yet this apparently timeless quality of Walden Pond is deceptive. The land surrounding Walden has been logged, farmed, and grazed since the earliest European settlements were founded more than three centuries ago. Walden Pond itself was repeatedly harvested for ice, which was sold around the world by Yankee merchants. And the forests now come to the edge of the water, having grown up over the past 160 years since agriculture was abandoned in the region.

Today scientists have identified an even more universal and widespread factor that is affecting the environment of Walden Pond and surrounding Concord: global warming and the broader impacts of climate change. Anthropogenic (human-caused) climate change has the potential to radically alter environments and ecosystems throughout the world, including Walden Pond. It turns out that Thoreau's observations made in the mid-nineteenth century can be used to measure the effects of a warming climate on the plants growing around the pond.

Thoreau as a Scientific Observer

Henry David Thoreau is best known today as a member of the transcendentalists, a pioneering environmental philosopher, an abolitionist, and an advocate of civil disobedience. In addition, he was a keen observer of changes in the seasons and differences in the landscape from one year to the next. *Walden* contains chapters devoted to individual seasons, and he intended to expand his later observations into a book entirely about the seasons (a project he never completed by the time of his death in 1862 at the age of forty-four).

During the 1850s, it's fair to say that Thoreau's love of observing natural phenomena became an obsession. As part of a decade-long project, he carefully recorded the timing of seasonal changes for every species of plant and animal that he knew in Concord. He had set his goal — he wanted to develop a calendar based on nature's patterns.

Thoreau walked in the woods throughout the year, observing and writing down what plants were in flower and in leaf, what plants were bearing fruits, and what the birds, frogs, fish, and other animals were doing. Based on these observations, his goal was to be able to tell exactly what day it was just by the nature he saw around him during his walk.

During this active period in his short life, Thoreau strolled around Concord for at least four hours almost every day, diligently noting the developments and events of nature. His effort was extraordinary, as evidenced by the extensive records he left behind in his field notes and journals. For eight years, he recorded the flowering times of over three hundred plant species, in addition to many observations of other seasonal events, such as the arrival of migrating birds, the leafing out of shrubs and trees, and ice-out on Walden Pond.

Thoreau worked on this calendar as a scientist, but in his typical style, he used his observations to address issues that crossed disciplines. Over the course of time, he extracted information on flowering times from his field notes and journals and compiled them into tables on large sheets of surveyor's paper — the action of a scientist. However, Thoreau was also preoccupied with the connections among beauty, philosophy, and the natural rhythms of the seasons. He wanted to use his time with nature to enrich his own sense of life. He wanted to see as deeply as possible *into* nature, not through it or beyond it.

I would argue that Thoreau was a climate change scientist whose research was more than a century ahead of its time; had his biological studies of the seasons been published and widely read, they likely would have had profound impacts on modern ecological thinking. Thoreau had observed that climate, particularly temperature and rainfall, affected the timing of the seasons in nature, including the appearance of flowers, the arrival of birds, the flight of insects, and the calling of frogs. His ideas influenced my thinking about the impacts of climate change, and his records of seasonal changes formed the starting point for my own research program.

Deciphering Thoreau's Journals

Once Abe Miller-Rushing and I had obtained a copy of Thoreau's list of flowering dates of Concord plants from the 1850s, the obvious course of action was to repeat his observations of the same plants today. Where Thoreau had observed the white bell-shaped flowers of the highbush blueberries on the shores of Walden Pond and the yellow buttercups in the moist meadows along the Concord River, Abe and I could do the same, later joined by numerous other students. I could see that Thoreau's observations from the 1850s were not just dead numbers, but rather a living link to the past — and a key to understanding the modern impact of global warming. When Thoreau described his ability to interpret patterns of water movement in the sand, he could just as easily have been describing my own excitement in discovering his tables of flowering times: "The earth is not a mere fragment of dead history, stratum upon stratum like leaves of a book, to be studied by geologists and antiquaries chiefly, but living poetry like the leaves of a tree, which precede flowers and fruit, — not a fossil earth, but a living earth" (*Walden*, 334).

Thoreau's tables list hundreds of species and their first flowering dates in specific years, from 1851 to 1858. The first year, 1851, is missing values for many species and thus was not very useful for our analysis; if I had to guess, I'd say this was probably because Thoreau was still learning where and when to find the first flowers of the spring for each species. Abe and I had the same difficulty in 2003, when we were still learning the paths, streams, and conservation areas of Concord. By the time we located many plant species, they were already in full flower or had even finished flowering.

One of the most difficult plant species to locate that summer was the fringed polygala, a low-growing perennial herb with small pink flowers with a feather fringe on the lower edge of its lower petal. After much searching, I found one small patch along the trail going through the old Rifle Range, where past generations of Massachusetts soldiers had practiced their shooting skills before being sent across the oceans to fight. The rusted mechanisms for raising and lowering the paper targets were still in place. By the time I had found the delicate polygala flowers, the plants were in full bloom, and the date when Thoreau had first observed this species was well past. My observation date for 2003 was thus of no use for our data set, but at least I knew where to look in 2004 and subsequent years for the first open polygala flowers.

THOREAU'S OBSERVATIONS THROUGH the middle years of 1852–56 were the most complete for a large number of species. Overall, the data set was astonishing in its detail and scientific value. This was the perfect data set for measuring the impact of climate change on plant flowering times. I could see that by linking our own observations to these years of Thoreau, we could make important scientific discoveries and attract public attention to this issue. In 2004 climate change was growing increasingly prominent not just as a topic of scientific discourse but as the subject of heated political debate. I could see that the value of our work would be to take this general topic of climate change and provide a specific example linked to Walden Pond and the life of Thoreau.

Thoreau was not always a perfectionist when it came to scientific observation. By 1857 and 1858, Thoreau's lists again had many missing values of flowering times. At this point, he had contracted tuberculosis, and as his health deteriorated he probably found it too exhausting to go out as frequently to search for the first flowers of his plants. Hikes to observe nature were exciting but also tiring, as Thoreau noted on September 13, 1852: "I have the habit of attention to such excess, that my senses get no rest — but suffer from a constant strain." Yet even with these gaps, we had an astonishingly detailed portrait of the flowering times of Concord's plants from 160 years ago. What would Thoreau's numbers reveal?

Decoding the Data

Thoreau's handwriting is notoriously difficult to read, and at the beginning the Boston University undergraduates working in my lab had a real challenge deciphering the records.

"What is this one, Professor Primack?" one of my undergrad research assistants would ask when I entered the lab. "Is it *Rbolna*? Or *Rlirlra*?"

I would stare at the name along with the students, rapidly going over all of the hundreds of plant names in the botanical index in my brain. What could it possibly be? And then a name pops out: *Rhodora*, the bog azalea with large pink-purple flowers arranged like a flock of birds in flight. That's it!

"And what is this one? *Babimy*? Or is it *Ralinny*?" another student asks, pointing to a different semi-decipherable scribble. Again, the wheels in my head spin and out comes *bayberry*, the aromatic shrub of sandy soils that yields the waxy berries from which candles can be made. And then we enter these names and dates of Thoreau's into our database.

Even without considering Thoreau's contributions, the flora of Concord can be tracked in more detail than possibly any area of the United States. Only a small percentage of the towns and cities of the United States have been carefully surveyed for the plants that they contain. For these places, botanists have combed the landscape over the course of one or more years to make up a list of the species occurring in the wild — too brief a survey to give a true picture of the community of plants as it changes over decades. Concord, however, has been the subject of five separate floras over the past 170 years, a distinction that can be claimed by perhaps no other location in the United States. (Botanists use the word *flora* to mean not just the assemblage of plant species from a specific area, but the book or publication that describes them.) No one factor explains why botanists have focused on Concord, but many of these individuals had connections with botany professors at Harvard University and, later, the New England Botanical Club.

Thoreau's interest in the botany of Concord is linked to a long tradition of collecting and identifying wild plants and making lists, dating back thousands of years to China, India, and ancient Rome. Long before Thoreau began his plant studies in the mid-nineteenth century, botany had been considered a suitable pastime for well-educated gentlemen and ladies. Thoreau took this long tradition of creating lists of plants one step further by making detailed observations of when the plants first flowered each year.

Starting in the late nineteenth century, the gentlemen intrigued by plants had the option of associating with the New England Botanical Club. (The club, founded in 1895, did not admit women as members until 1964.) For over one hundred years, members of this small group of plant enthusiasts have been seeking out rare plants across New England in the spring and summer, collecting herbarium specimens, and spending the long winter season comparing notes. Such scientific connections have been and still are useful in providing botanical students with training in plant identification and access to the latest scientific findings. The herbarium specimens collected by club members consist of dried, flattened plants attached to heavy paper sheets along with labels that list the date and place of collection and the collector's name. These are currently housed at the Harvard University Herbaria in Cambridge.

Our Fieldwork Begins

Thoreau took quite seriously his task of recording the first flowering dates of plants and the first signs of animals. And as a lifelong observer of nature, he took pride in his own ability to find the first flower of the season, noting, "It will take you half a lifetime to find out where to look for the earliest flower" (*Journal*, April 2, 1856). He described finding rock crevices where the warm rocks trap sunlight to create a natural greenhouse that stimulates early plant growth and flowering in the spring. After many years of studying every corner and out-of-the-way spot of Concord, Thoreau knew where to look for the first blooming marsh marigolds and bluets. And he searched such places diligently to find the *first* flower of the year. "I often visited a particular plant four or five miles distant, half a dozen times within a fortnight, that I might know exactly when it opened" (*Journal*, December 4, 1856).

In 2003 I also began to search for the first flowers of spring in Concord. As I mentioned earlier, that first year I often missed the first flowers, because I did not know where to find the plants. When I found a new kind of plant for the first time, it was often in full flower or even past flowering. But for the next three years, Abe and I became increasingly obsessed with searching throughout Concord to find the first flowers of each species, work that has continued every year right up to the present. The one exception was 2007, the fifth year of the study. At this point, Abe had left Boston for a postdoctoral position, and I was so uncertain about my future research direction, that I did not make regular observations. I missed the first flowering data of so many species that the data was not usable. By 2008 I would regain my sense of purpose and enthusiasm for fieldwork.

FOR MOST YEARS, I traced Thoreau's footsteps, walking extra miles on the first warm days of spring, searching out hundreds and even thousands of plants, to find that first open flower. On some of the early days of late March and early April, I walked across long stretches of snow-covered ground on south-facing slopes looking for patches of bare ground with early spring growth. I examined the northern edge of Walden Pond, where the curved south-facing sandy banks form natural parabolas, focusing the sun's energy on the early lowbush blueberries and fire cherries that line the adjacent edge of the pond. At such especially warm places, I would find the first blueberry flowers. I would not spend much time looking for first flowers on the western banks. Even though the western bank might catch the rising sun in the east, it would quickly become shaded and cool by midmorning: not a place to find an early flower.

Another favorite spot where I searched for plants that required disturbed, open ground was the narrow path between a railroad track embankment and the industrial buildings of West Concord, where the heat of the sun accumulates on still, sunny days, creating an oven-like effect. That was a good place to find the first bladder campion flowers, with their swollen calyx below the narrow white petals. The earliest swamp plants, such as silky dogwood and the fragrant white-flowered swamp azalea, I would find on a narrow road that cuts through a marsh leading to the Middlesex School, an elegant private high school. In that spot, the sun hits the edge of the raised roadbed and warms the air and water, causing the plants to flower a few days earlier than anywhere else in Concord. Following in the footsteps of Thoreau, I learned these spots and visited them regularly each spring. This may seem like "cheating," but it's not — Thoreau had also found by long experience the few places where either natural circumstances or human modification of the environment had created a warmer, and often a drier and sunnier, place where each species flowers earlier in the spring.

To make his observations, Thoreau learned to travel light, allowing him to walk all day and not get tired. His field equipment consisted of a music book with which to press plant specimens flat, a pencil, a notebook, a telescope for watching birds, a pocketknife, and a hand lens. He also sometimes carried plants in a special straw hat with a built-in shelf. He was particularly proud of his highly functional hat: "Some whom I visited were evidently surprised at its dilapidated look, as I deposited it on their front entry table. I assured them it was not so much my hat as my botany-box" (*Journal*, December 4, 1856).

My own field equipment was not so different — a pencil or ballpoint pen, a notebook, binoculars, a digital camera for recording observations, and a field guide for wildflower identifications. After the first two years, Abe and I also carried cell phones to coordinate the places we visited and the time and place we would rendezvous at the day's end. Even if cell phones had been available in Thoreau's time, he may not have wanted one, as he preferred to walk alone when he was making serious observations. He recognized that when he walked around with friends, he tended to talk and not observe nature very closely.

During the academic year of 2003–4, when the ground was covered in snow, we scoured both Thoreau's table of flowering times and his journal entries to compile a list of 578 plant species (both native and nonnative) that Thoreau had recorded as occurring in the wild in Concord. We searched for these plants during the next three years, walking through Concord one to three days per week, from April to September, deliberately seeking out places where rare plants might occur.

Our fieldwork plan sounded simple enough on the face of it. Go to Concord; scope out places where the wildflowers observed by Thoreau might exist; survey these locations repeatedly and record what we saw in bloom on a given date; go back and do it again a year later, and again the year after that. But, of course, fieldwork is never that easy. One major challenge was just learning to identify all of Concord's plants. Two of our regular assistants were premed students who knew virtually nothing about plants; we had to teach them to identify wildflowers. For another, walking all over Concord looking for wildflowers sounds easy enough until you try it. It's actually arduous work. Concord covers a lot of ground at twenty-five square miles. And there was another obstacle, one that we didn't realize might be there until we were well into our fieldwork: many of the plants Thoreau and later observers had recorded simply were no longer there.

*Pink lady's slipper orchid flower; photo by Richard B. Primack
and Abraham J. Miller-Rushing.*

*How long some very conspicuous ones [wildflowers]
may escape the most diligent walker, if you do not chance
to visit their localities the right week or fortnight.*

THOREAU, *JOURNAL*, MAY 1853

4. Phantom Plants

WE GRADUALLY REALIZED that there were a lot of species that Thoreau and later botanists had recorded in Concord that we were not seeing. Of the species seen by Thoreau in the mid-nineteenth century, mainly the 1850s, we failed to find fully a quarter of them during our first three years of searching in 2003, 2004, and 2005 — and no one that we spoke with seemed to be able to tell us where to find them. These are species that we call locally extinct in Concord, though they all occur elsewhere in Massachusetts, often in neighboring towns. And for many other species, the plants were still present in Concord but in very limited numbers. We found that a further third of the plant species seen by Thoreau, above and beyond the ones we hadn't found, were present in Concord in only one or two populations, sometimes quite small populations at that.

Depending on one's perspective, this result is either encouraging or discouraging. If you want to take the glass-half-full point of view, it's amazing that even after 160 years, three-quarters of the species that Thoreau observed in Concord can still be found there. On the other hand, the half-empty perspective notes that even with the extensive numbers of parks and other protected lands found in Concord, a quarter of the species Thoreau saw are botanical phantoms, plants that have vanished from the landscape. And since rare, native species are most likely to go extinct, and a lot of the smaller populations we recorded were both rare *and* native, we could speculate that within a few decades, if nothing is done to prevent it, those species we saw in only a few populations — a third of Thoreau's original list, remember — could also disappear. That would mean that around half of the species that Thoreau observed in Concord will no longer be present in a few decades from now; they are destined for local extinction.

In our investigations of Concord, we recorded 98 species as occurring in only one place. A further 24 species are what we consider to be very rare, with populations of only 2 to 10 individuals. And 9 species are extremely rare with just a single individual. These extremely rare species include formerly common species that are down to their last individual, such as the Canada lily. Species like the Canada lily formerly occurred in abundance in the river meadows of Concord, thriving in sunny, moist habitats. However, now the river meadows are overgrown with trees, shading out the plants; deer graze along the river, eating the tender shoots of the lily; and aggressive invasive species, such as purple loosestrife, crowd them out of their remaining habitat. A species that is down to its last few individuals is in serious danger of local extinction, and a species that now exists as one remaining plant is all but extinct, unless a major effort can be made to protect and manage it back to health.

It's possible, of course, that we may have missed recording some plant species if they did not flower during our ten years of observation, in which case we might have overlooked them and therefore not counted them. This may be particularly true for orchids with narrow grass-like leaves, which can be inconspicuous and hard to spot when not in flower. Orchids may even remain dormant in the ground for many years, only to reappear when conditions are favorable. However, we believe that the number of such overlooked species is small. For example, the false hellebore, a flower in the lily family, occurs in a swampy area of the Estabrook Woods, but in our many years of fieldwork it never flowered. Yet we were able to recognize this species by its broad, pleated leaves and overall similar appearance to a cornstalk. So we would like to think that we spotted most of the plants we were looking for and didn't overlook any. Even if we did, though, the overall decline in plant species is clear.

The number of plant species growing in the wild and their abundance are declining over time in ways that can be quite dramatic and localized. In April 2013 I noticed with considerable disappointment that the one trailing arbutus plant growing on the trail around Walden Pond was dead. Only the dried brown leaves remained of this beautiful evergreen wildflower species with pink flowers that is also the Massachusetts state flower. This plant was the only representative of its species that I ever saw on public land in Concord and one I had shown to hundreds of students and adults on walks around the pond. Thoreau records it as common in many places, and the species still occurs in two Concord localities that are off-limits to the public. However, where will people in Concord now be able to see the trailing arbutus?

The river meadows along the Sudbury, Concord, and Assabet Rivers need to be considered in the most detail, because these are the places where there used to be such a cornucopia of wildflowers. The Sudbury River, flowing into Concord from the south, and the Assabet River, flowering into Concord from the west, merge together in the center of Concord to form the larger Concord River, which flows northward for several miles, under the Old North Bridge, before leaving the town. These wet fields along the rivers were the very reason that the European settlers were attracted to Concord in the first place, but the meadows are considerably altered from Thoreau's time. In the past, river meadows would flood in the spring when the water levels were high from the melting snow upriver. During the spring and summer as the water levels declined, grasses would grow thickly in the sunny, moist meadows. At the end of the summer, the meadows were mown by the farmers for hay, which kept them open for late summer and fall wildflowers. Today the meadows are no longer mowed for hay, as there are so few horses, cattle, and sheep left in Concord. It is now also cheaper to import hay for livestock for the few remaining farms. Many abandoned meadows have been colonized by the seeds of nearby trees, especially maples, and have now become tree-filled swamp forests.

This process, called succession, was well described by Thoreau, who was one of the first people to carefully investigate it. Where river meadow habitats do remain, they are often dominated by stands of exotic species, such as purple loosestrife. Either way, native river meadow species now have few areas in which to grow in Concord. The only places where many river meadow species can still be found are in the Old Calf Pasture near the center of Concord and around the Old North Bridge at the Minute Man National Historical Park, where land managers actively mow meadows and remove trees. In these places, the sole remaining populations of certain species persist, such as a single population of the green-fringed orchid at the Old Calf Pasture.

Revolutionary Viewpoint

Eight years ago officials at the Minute Man National Historical Park became concerned that too many silver maple trees and red maples trees had taken over what was formerly river meadow near the Old North Bridge along the Concord River. This site is one of the key battlefields of the Revolutionary War where the American militia, known as the minutemen, rallied and forced the British redcoats back to Boston. Over time a tree-filled swamp came to dominate the view of the bridge from the visitors center, though at the time of the war the whole area along the river would have been fields and pastures.

Park officials decided to restore the Revolutionary War appearance of the site by removing the swamp trees on the riverbanks near the bridge. According to what I have been told, before carrying out this plan, they dutifully informed the Concord town officials of their intention to cut down the trees, as the area is a wetland and governed by special Massachusetts regulations. To the surprise and dismay of the park officials, the town's environmental officer insisted that they needed the town's permission before any trees could be cut down. And when certain local environmental groups and citizens heard about the plan, they argued that any cutting of park trees was wrong. After several public meetings, the park officials became exasperated with a process that was blocking their management plan and potentially putting them, a federal agency, under the jurisdiction of a town government. So the park officials simply informed the town that they were going ahead with the plans to cut the trees anyway.

This is where the story becomes murky. According to one park official who talked at length with me about the incident just after it happened, the town officials responded that they would send the local police to arrest anyone cutting trees along the river, as this was in direct violation of state laws protecting wetlands. The park officials countered that they would call in soldiers, if necessary, to carry out the tree cutting! Since many of the key players in the incident are no longer active in Concord, it is hard to know if these were real threats or were just made in jest. In the end, the town officials took no steps to block the removal of trees.

At present, the relationship between the park and town officials is much more cordial than it was eight years ago, and park officials minimize this past conflict. It is hard to know the reality of this incident and what was just bluster. In any case, this illustrates the difficulty of satisfying the conflicting demands of conservation management — in this case the need to restore a historical landscape and the requirement to maintain vegetation along rivers.

Removing the trees at the Minute Man National Historical Park has allowed some native plants to return to the meadow. In the summer of 2005, when we were searching for early flowering times, we recognized that the newly cleared river meadow at the Old North Bridge was an important place to check due to the large expanse of open, sunny habitat. One day while pushing through the thickets of knotweed along the riverbank, I spotted the characteristic large orange hanging flowers of the Canada lily emerging from the middle of a dense clump of plants. The flowers were produced on a tall stem that reached above my own height, nearly six feet tall. This was the first wild lily plant that we had seen in all of our time in Concord.

The large orange flowers, each one with six petals curved backward, were a rare treat to behold. We had thought that lilies had disappeared from Concord. By restoring the river meadow at the Old North Bridge, the park service had provided new habitat for this formerly common lily. As we walked back to the parking lot, we happened to pass Chris Davis, the environmental officer for the park and one of the people who had led the park's efforts to remove the trees. When we told him about this lily plant, at first he did not appear that interested. However, once we explained its significance as an indicator of how tree removal had encouraged the appearance of a locally rare wildflower species, he quickly got the point. He became excited by our discovery and headed down to the meadow to photograph the lily.

"Terror" on the Tracks

In the course of the project, I became dedicated, perhaps even obsessed just like Thoreau, with finding the first open flowers of each species in the spring. I learned that I could often find the earliest flowers of certain species of open habitats, such as the ovate-leaved violet, growing among the gravel placed along the railroad bed of the commuter line just above

Walden Pond. This railroad is the same line that Thoreau saw and wrote about — it was within sight of his cabin, and he would wave to the conductor as the train passed. The railroad bed is exposed to the direct rays of the sun all day, and the dark gravel gets progressively hotter throughout the day. The heat can be uncomfortable during the summer.

The tracks could also become uncomfortable in other ways, as I discovered on May 5, 2005. That day I was carefully searching the gravel bed, trying to find the first flower of this pleasant springtime violet, with its egg-shaped leaves and blue- to lavender-colored flowers. At first, it seemed that none of the flower buds had opened yet. Finally, after some minutes I found a few open flowers on a plant growing among the gravel. I recorded my observation in my field book and then decided to photograph the flower to document its occurrence. I got my camera ready, but the intense direct sunlight on the gravel bed was too bright. So I spent some minutes rigging up some shading with my field guides to create the best possible picture.

While focusing my eyes downward and concentrating on my violet photography, I kept my ears cocked for the sound of an approaching train behind me on the tracks. As it turns out, I should have kept looking up as well. A heavy construction truck riding almost silently on the rails pulled up to within twenty yards of where I was kneeling with my camera. When I heard the sound of the truck, I jerked up with a real sense of fear, not knowing what the sound was. For all I knew, a train was about to clobber me. The men in the construction crew were all staring at me, trying to figure out what in the world I could possibly be doing.

Needless to say, I was feeling awkward and embarrassed. I felt like running into the woods, but I knew it would be a bad idea, and so I stood frozen in place. Eventually, the driver of the truck came over to me and angrily demanded to know what I was doing there. When I told him that I was a botanist and photographing wildflowers, he replied, "How do I know that is true? Don't you think a guy with a camera on the railroad tracks is suspicious?" When I offered to show him my photographs of violets, he showed no interest but told me that I was on private property, and that he would report me to the police and the FBI. I told him that I understood that I was trespassing on private property, and I would not be on the tracks again — and the driver finally just walked away disdainfully.

I knew that walking on the railroad tracks presented the potential danger of being hit by a speeding train. Now I knew that they were also being carefully watched due to post-9/11 security concerns. And I wondered whether the driver really thought that a mild-mannered university professor photographing wildflowers posed a terrorist threat. Given how Thoreau felt about public authorities (his arrest for failing to pay his poll tax was the genesis of his famous essay *On Civil Disobedience*), he likely would have encouraged us to keep at it, and indeed we did. Abe and I continued to monitor the flowering times of plants along the tracks, but we kept our times there as brief as possible, and we kept looking in both directions on the tracks for trains, police, and railway crews. We had a few more run-ins with police patrolling the tracks on horseback, which kept us on our toes.

Pink Lady's Slipper Orchid: A Sign of Warmer Times

In 1852, as he was embarking on his second year of systematically observing flowering times, Thoreau was making extremely detailed observations of birds and flowers. On May 28, he recorded the first open flower of the pink lady's slipper orchid. This is a distinctive wildflower of open woods; it occurs in many places in Concord. Some plants, like white oak trees and paper birch, have tiny flowers. Even a botanist has trouble determining the exact date when these tiny flowers are truly open. But the opening of the pink lady's slipper orchid flower is unmistakable. Over the course of a week, the flower bud is elevated a foot above ground on a green stalk. The immature flower bud gradually changes from green to pale pink. Finally, when it is ready to open, the flower pouch puffs out and turns bright pink and the petals flare out to the sides.

This orchid species is a favorite among naturalists, as it is among the largest of the orchids and fairly easy to find. The pink lady's slipper orchid is also an easy plant to recognize both in flower and fruit, and one that Thoreau knew well. In 1853 he noted the first flowering of this species on May 20, and between May 24 and May 30 in subsequent years. For Thoreau, this was a late May species.

Today, if I went looking for the first flowers of the pink lady's slipper orchid on May 20, I would be too late — some plants might still have fully opened flowers, while others would have only wilted flowers or even the beginnings of young green fruits. In my own wanderings around Concord in 2004, I found the first flowering of the pink lady's slipper orchid on May 14, about ten days earlier than a typical Thoreau year, and this was no fluke of an abnormal year. In both 2005 and 2006, this species started to flower on May 17. During the incredibly warm spring of 2010, this orchid actually began to flower on May 4. At the time, I thought this would be the earliest I would ever see the species flowering in my lifetime. However, in 2012, after an exceptionally warm and almost snow-free January, February, and March, the pink lady's slipper orchid flowered on May 3. In the 160 years since Thoreau recorded his observations on scraps of paper during his daily walks, the pink lady's slipper orchid in Concord has shifted its first flowering time so that it now flowers three weeks earlier than in the past. This change happened gradually, so that only by comparing records taken at widely spaced intervals — Thoreau's records in the 1850s and our own observations made 160 years later — could I see the alteration in flowering times over the intervening decades.

The pink lady's slipper isn't the only species that shifted its flowering time. Other species have also changed, particularly species that flower in the springtime. Consider apple trees: when apple trees flower, they are covered with large five-petaled blossoms that range in color from white to pink. Wild apple is a very conspicuous species that Thoreau would readily see in his walks along forest edges, and that modern Concordians driving in their cars can spot from the roadside. Thoreau was especially fond of wild apples and wrote that they had greater value than the exotic fruits, such as pineapples and oranges, that were brought into Boston from tropical regions by ship. He enjoyed picking up wild apples from the ground and eating them during his walks. He did not mind if mice and other forest creatures had nibbled them already. And today I also enjoy eating apples for my lunch, although these come from apple trees grown thousands of miles away or even across the ocean. In the 1850s, Thoreau recorded apple trees as first flowering between May 14 and 21, but in 2004, 2005, and 2006, apple trees flowered on May 7, April 27, and April 29, respectively, along Concord roadsides, where the streets absorb the sunlight and create an especially warm location. And in 2012, apples flowered on April 18.

Apple trees have shifted their flowering times in Concord, now flowering an average of two to four weeks earlier.

Or consider the yellow wood sorrel (*Oxalis europaea*), a small herb that grows in fields and roadsides; this common species is quite distinctive, with five-petaled yellow flowers and three-parted leaves that look like clover. Because it is a small plant of disturbed habitats, many naturalists would not want to bother with it. People interested in edible wild plants know wood sorrel because of the pleasant, acidic taste of the leaves, somewhat like spinach. But it is another key species in detecting the signature of climate change. Thoreau recorded this species as first flowering at the end of May or early June during the 1850s, but in recent years this species can be found in flower as early as April 22, almost six weeks earlier than in Thoreau's time. The earliest place in Concord where this species flowers is in loose gravel right next to the south-facing edge of the library of the Middlesex School. At this location, a hot spot is created on sunny March and April days by the sun hitting the bottom edge of the building where it meets the gravel, allowing the plants to flourish and then flower even when the air temperature of the surrounding area is still cold.

For a whole series of species, we found that plants now flower earlier in the year. Not all species shifted the same amount: some changed by one week, others by two weeks or more, and some species did not change at all. But the general shift toward earlier flowering is a widespread pattern. Thoreau's obsession with finding a natural calendar to tell the season during the 1850s has in fact provided us with a biological yardstick to measure precisely how much the changing climate has affected plants growing in the Concord area over the past 160 years. Things are clearly changing. But is it global warming?

Mercury Rising

Part of the difficulty in deciding whether the changes we are observing are due to a warming climate or something else is that we seemed to have no information about what happened during the intervening century and a half. Yes, our data showed multiple species flowering earlier than they had in Thoreau's time, but either our readings or Thoreau's could be simply an aberration. Without knowing anything about the intervening century-plus, it is difficult to say.

Fortunately for us, Thoreau was only the first of a number of Concord naturalists who continued to record the timing of natural transitions, inspired by his example. One of the most important of these people for my students and me was a botanist named Alfred Winslow Hosmer.

Thoreau's Mantle Passes to Hosmer

Alfred Winslow Hosmer was a shopkeeper and a photographer, in addition to being a keen amateur botanist and a great admirer of Thoreau. Beginning in 1878, he recorded the first flowering date of every plant in Concord that he could find. Following a nine-year break, he recorded first flowering dates in Concord every year from 1888 until 1902, near the end of his life. During this intense period, he recorded the flowering dates of over seven hundred plant species and copied all of his observations into a beautiful storekeeper's ledger that is stored today in the Special Collections section of the Concord Free Public Library.

Based on his field records, Hosmer made observations several days a week from March to October, traveling all over Concord on foot. He even turned over the leaves of the dozens of species of ferns in Concord, recording when they released their spores. Like Thoreau before him, Hosmer spent an extraordinary amount of time examining the flora of Concord. In contrast to Thoreau, Hosmer never wrote about this massive undertaking, so we don't know why he made these extensive observations of flowering times, or if he reached any conclusions. My best guess is that Hosmer wanted to continue the work out of admiration for Thoreau and Thoreau's botanical work. Hosmer's friend Samuel Jones concurred that Hosmer was following in Thoreau's footsteps, when he wrote in a letter: "Fred is . . . better informed about Thoreau's haunts than any man living or dead. I, poor miserable I, admire Thoreau; Fred lives him!" (*Toward the Making of Thoreau's Modern Reputation*, 288). It is also possible that Hosmer just loved botanizing, and that was enough of a reason to take up Thoreau's work.

When we entered Hosmer's observations into our analysis, we found that his dates of first flowering were exactly intermediate between Thoreau's late flowering dates and our more recent early flowering dates. So plants were changing their flowering times and becoming earlier with each passing decade.

Once I learned about and located Hosmer's work, my students and I had two sets of historical observations to measure when plants flowered in the past: the first set made by Thoreau from 1851 to 1858, and the second set created by Hosmer thirty to forty years later, in 1878 and from 1888 to 1902. To these data sets, we added our own observations: in 2004–6 and again in 2008–12, Abe and I hiked through Concord making the same observations on the same species at many of the same localities as Thoreau and Hosmer. They had skirted the edge of Walden Pond, looking for the first white shadbush flowers, with their long white petals, standing out against the greenish-blue of the water. A century and a half later, we had done the same. Despite the passage of time, the plants were mostly still in the places where Thoreau and Hosmer saw them, although some of their favorite spots were now covered with parking lots, houses, or highways.

With these observations in Concord, Abe and I analyzed the flowering times of the forty-three most common plant species that Thoreau, Hosmer, and we saw in each year — species such as bluets, white clover, marsh marigold, and highbush blueberry. By using all of these species at once, we could generalize about the impacts of climate change, yet avoid the problems that arise when a particular species is not representative, or when a species' flowering is altered in some years by factors other than temperature, such as flooding, disease, or overbrowsing by deer. When Abe and I calculated the average flowering time for these common species in a given year, we found that in Thoreau's time, these species flowered on May 16. In Hosmer's years during the end of the nineteenth century, these same species flowered on May 12, and during our first three study years of 2004–6, these plants flowered on average around May 8. Our more recent observations of 2008–12 — and especially the recording-breaking warm years of 2009, 2010, and 2012 — are even earlier, with a new average date of observations of May 5, seven days earlier than in Hosmer's time and eleven days earlier than in Thoreau's time. In fact, the flowering times for 2009, 2010, and 2012 are so early that they are "off the charts." The plants had an average flowering time of April 30 in 2009, and the flowering date had shifted to April 21 in 2010. Flowering times for 2012 were even earlier, averaging April 20. So in these three unusually warm recent years, these forty-three common plant species were flowering three full weeks earlier than in Thoreau's time. But this was the *average* flowering time across all forty-three species. Some individual species, like the high-bush blueberry and the apple trees that we mentioned earlier, are now flowering four or more weeks earlier. The flowering times of other species, such as the graceful white-flowered early saxifrage and pink-flowered nodding trillium, are barely changing at all.

Heat Islands

What could explain this shift in flowering times? Why were some species now flowering three weeks earlier than they did in Thoreau's time? Of all of the possible factors that might have caused this shift, one factor was paramount: the climate in eastern Massachusetts has warmed since Thoreau's time, and plants are responding to the temperature increase. At Blue Hill Meteorological Observatory, on a hill southwest of Boston, diligent meteorologists have collected weather data for almost two hundred years. They have tracked the weather for a longer time in a single location than any other weather station in the United States. During the 1850s, the average yearly temperature was around 46 degrees Fahrenheit. By Hosmer's time (the 1890s), Boston temperatures reached an annual average of 48 degrees, and by our time, temperatures had reached 50 degrees.

These temperature increases are driven mainly by the increasing development of the Boston metropolitan area. People have built ever more buildings, parking lots, streets, and other dark paved surfaces that absorb the energy of the sun and store it as heat. These hot surfaces have replaced trees, which reflect the sunlight and release water from their leaves, cooling the environment. Pavement, buildings, and the "concrete jungle" of a city create a bubble of warm air that covers the urban landscape in a phenomenon that scientists call the *urban heat island effect*. In the case of eastern Massachusetts, this bubble of warm air is centered on downtown Boston and currently extends just beyond Concord into the far western suburbs.

The urban heat island effect accounts for about two-thirds of the increase in Concord's and Boston's temperature since the 1850s. The other third is attributed to general global warming of just over 1 degree Fahrenheit that has affected all regions of the world. So even though many Concord residents see themselves as living apart from metropolitan Boston and living in a way that more closely harkens back to traditional Yankee values of self-reliance — with more democracy, more appreciation of ideas, and more respect for individuality and individual responsibility — when it comes to climate, Concord is firmly within the influence of Boston.

The increase in temperature that eastern Massachusetts has already experienced due to global warming and the urban heat island effect is almost the same amount that the rest of the United States is expected to warm during the next eight to nine decades of this century as a result of just global warming. As a consequence, environmental studies in the metropolitan Boston area and in other large northern cities such as New York and Chicago have the potential to demonstrate right now the types of impacts that climate change will have later this century in the rest of the country. Concord is a living laboratory for the effects of a warming climate on species diversity. If you're wondering if you can take steps to reduce the effects of climate change where you live, you can. We'll get to that later, along with more general steps for individuals and societies to reduce their production of greenhouse gases.

A warmer Concord is even evident in the attire worn by Thoreau compared to what my students and I wear today. Thoreau's field clothing was suited to rough walking in cold weather: heavy shoes, old corduroy clothing, and his "botany box" hat. In contrast, my field clothing consisted of a long-sleeved cotton shirt, lightweight pants, a baseball cap, and running shoes. A windbreaker for rare cooler days completed my wardrobe. This lighter style of clothing reflects the warmer climate since the time of Thoreau in combination with improved technology. In fact, many of our sunny field days in the summer were cut short due to extreme heat. We would be so tired, thirsty, and overheated from a long day of hiking that it was unpleasant to continue.

Plants Respond to Temperature

Thoreau was aware that plants responded strongly to warm spring temperature. On March 30, 1853, he observed:

I am not sure my willow will bloom fairly to-day. How warily the flowers open! Not to be caught out too early, not bursting into bloom with the first genial heat, but holding back as if foreseeing the transient checks, and yielding only to the absolute progress of the season. However, some hardy flowers which are quite ready will open just before a cold snap, while others, which were almost equally advanced, may be retarded a week.

When we looked at the same forty-three common spring-flowering species of Concord in relation to late winter and spring temperatures, we saw that plants flowered later in years with cold spring temperatures and earlier in years with warm spring temperatures. We were able to describe this relationship precisely: plants flowered two days earlier for each 1 degree Fahrenheit increase in temperature. With average temperatures predicted to rise by as much as 9 additional degrees Fahrenheit in the coming century, we expect that plants will continue to flower even earlier, perhaps as many as eighteen days earlier than they do now. And if there is a year that is warmer than average, plants might flower a month earlier.

We also used statistical tests to simultaneously analyze the effects of temperature and the passage of time on plant flowering times. These tests showed us that the earlier flowering of spring wildflowers in Concord over the past century and a half is due almost entirely to warming temperatures in the spring. All we need to know are the average temperatures in the months of March, April, and May. We do not need any other variables to explain spring-flowering times. Considering rainfall, humidity, wind speed, and snow cover does not give us any additional explanatory power over just using air temperature. We don't even need to know how cold it was in the previous winter or autumn. Plants are flowering earlier in successive years because the spring temperatures are getting warmer.

Yet, in looking at the entire list of plants from Concord, we saw that not all plants shifted their flowering times in response to a warming climate. The plants that responded to warming temperatures were mostly those plants that flower in the spring, plants like marsh marigold and black birch. These species form their flower buds during the end of the previous summer; the buds wait through the fall and winter until conditions become warm enough for the ground to thaw and the sap to flow. Then the buds swell and open. Such spring-flowering plants contrast with summer-flowering plants — such as mints, goldenrods, and asters — which form their buds over the course of the spring and summer. Their flower buds never see winter. For the most part, these summer-flowering plants did not change their flowering times in response to temperature; they probably timed their flowering to the number of hours of daylight and perhaps to the amount of rainfall and nutrients in the soil. So we found that spring-flowering plants are the best indicators of the effects of global warming.

Timing Matters

In many passages in his journals, Thoreau observed that the natural world responds to changes in climate with the arrival of spring. But although each species seems to respond on its own, many relationships between species rely on timing. When bees become active in the spring, they need flowers to provide nectar, and the flowers need bees to pollinate them. When chipmunks, mice, and other herbivores emerge in the spring, they need the snow to have melted and plants to be growing and ready to eat. When birds arrive in New England from their wintering grounds in the tropics, they need insects to eat. If the flowers, plants, and insects are not around when bees, mammals, and birds need them, those bees, mammals, and birds can go hungry. Not only would they risk starvation, but many plants and animals that depend on them would be in trouble too. Nature is a web of relationships, in which every change in one plant or animal species has consequences for many other plants or animal species.

Let's consider a hypothetical case, just a simple food web consisting of a red maple tree, a common violet plant, a wild bee, a caterpillar, and a Baltimore oriole. The caterpillar emerges in the spring just at the same time that the leaves unfurl on the tree, providing tasty and nutritious young leaves for the growing larva. As the leaves mature, they develop toxins that slow the growth of the caterpillar. The oriole winters in Ecuador, arriving in New England in the spring to breed. Its eggs typically hatch at the same time as when the caterpillars are at their most abundant, thus providing plenty of food for the growing chicks. The violet plant flowers and the wild bee emerge simultaneously, which is good for both of them because the bee pollinates the violet flowers. It so happens that the bee pollinates the violet a short time after the tree leafs out, although the violet and the wild bee generally do not interact with the tree, the caterpillar, or the bird.

What might happen if global warming messes with the timing in our neat little food web? The tree now leafs out earlier than it used to, and the caterpillar matches the change, as they are both responding to same *local* changes in temperature. But our bird misses the boat — from Ecuador, it can't tell whether it will be a warm spring or a cold spring in its northern breeding territories. It continues to migrate and breed at the same time each year, using length of daylight as its cue for departure for the trip north. When it finally arrives, the bird breeds too late to feed the caterpillars to its chicks. The oriole chicks could die of starvation unless the parent bird can feed them something else. Let's suppose our particular bird is flexible and learns to eat a multitude of other insects that happen to be around when it breeds. One of these unfortunate insects is our bee. The bee population declines dramatically because the bird is now eating it, whereas before it was mostly free from predation. As a result, there are not enough bees to pollinate the violet flowers. Consequently, the violets don't produce any seeds, and the population begins to decline. Meanwhile, the caterpillar is doing splendidly, eating lots of tree leaves.

One of the major worries about global warming is that it could disrupt relationships among species that have evolved over long periods of time. We saw the potential for this disruption in our own data. Some plants responded dramatically to warming, others less so, and others not at all. Time-sensitive relationships will change. We saw that plant relationships will change, but should we expect animals to behave any differently? In a later chapter, we will show that birds are also changing in their timing, but at a different rate than plants.

Thoreau's Records and Today's Climate

In the end, our historical records — Thoreau's and Hosmer's field notes — combined with our modern observations tell us that plants are responding to the warming conditions in the Concord area, as shown by earlier flowering and leafing out — and, moreover, this response has been occurring for quite some time. These increases in Concord's temperature over the past 160 years are due to human modification of the environment, including the building up of Boston and the associated urban heat island effect as well as the production of greenhouse gases causing global warming. Thoreau was aware of the immense human toll on nature that resulted from farming, logging, and other activities. Yet even Thoreau would have been astounded that human activity could affect the very temperature of the world and the onset of spring.

IN THE PAST, Thoreau directly called our attention to the issues of protecting nature, ending slavery and unjust war, and the need for simple living. Today his journals and his unfinished calendar of nature can give us further insights. His records of plant flowering times at Walden Pond and in one small town in Massachusetts convey a much larger truth. The changing climate is already affecting the plant life that forms the base of the food web upon which all life depends. Of course, natural processes other than food webs depend on timing too. Plants may grow more in years when they leaf out earlier and keep their leaves on later in the autumn. And when plants grow more, they absorb more carbon dioxide from the atmosphere and move more water from the ground to the air, altering global carbon and water cycles. Huge numbers of natural processes, ranging from the level of individual plants and animals to the level of global cycles, will alter as timing changes in natural systems.

As this example of flowering times shows us, the effects of climate change are not just happening in the Arctic or on a remote tropical mountainside. We can see these changes right around us, if we take the time to look. However, it may not be just plant flowering times that are changing in Concord and eastern Massachusetts. Birds and insects may be altering their activity patterns as well, threatening the ecological relationships that link species together. And such changing relationships may harm certain rare species, driving them toward extinction on a local scale.

*Spray of apple blossoms; Harold Wendell Gleason photo
1901.33 taken May 30, 1901, in old Estabrook Orchard, Concord,
Mass. Special Collections, Concord Free Public Library.*

The era of the Wild Apple will soon be past.
It is a fruit which will probably become extinct in New England. . . .
I fear that he who walks over these fields a century hence will
not know the pleasure of knocking off wild apples.

ATLANTIC MONTHLY, NOVEMBER 1862

5. Wild Apples and Other Missing Flowers

WHAT THOREAU PREDICTED has come to pass. In my walks around Concord, I have only rarely met wild apple trees, and even then they were too shaded to bear fruit. Other plant species have shared a similar fate.

On his walks, Thoreau recorded the first open flowers of the fringed gentian. This is a distinctive species, easy to identify, with thick, opposite leaves and a graceful four-petaled blue flower with long fringes on the petals. Botanists are always interested in gentians, because of their unique appearance and reputed medicinal value as an antifungal. In his journal of October 19, 1852, Thoreau records his observation of this late-blooming gentian:

> It is too remarkable a flower not to be sought out and admired each year, however rare. It is one of the errands of the walker, as well as the bees, for it yields him a more celestial nectar still. It is a very singular and agreeable surprise to come upon this conspicuous and handsome and . . . blue flower at this season when flowers have passed out of our minds and memories.

However, even though later botanists have recorded this species as common in Concord a century ago and occurring occasionally forty years ago, we have never located even one gentian plant in our years of searching. Why isn't it in Concord anymore? We had already noticed the "phantom plants" — plant species that Thoreau and other naturalists had noted from Concord that were now missing. Now we wanted to delve deeper and find out why.

From Common to Rare to Vanished

When you consider the fact that a species that was common in Thoreau's time could vanish from Concord, how much more likely is it that uncommon or rare species would have disappeared? One species that Thoreau perceived to be threatened by human activities is the smaller purple-fringed orchis, a dramatically beautiful orchid with a spray of intricate purple flowers on an erect stalk. On July 26, 1853, Thoreau noted in his journal that the "smaller purple-fringed orchis has not quite filled out its spike. What a surprise to detect under a dark, damp, cavernous copse [thicket of small trees], where some wild beast might fitly prowl, this splendid flower, silently standing with all its eyes on you." Yet even in Thoreau's day, this flower was becoming less common in Concord. On July 30, 1853, just a few days after his initial entry, Thoreau lamented the decline of this species, writing:

> In every meadow you see far and near the lumbering hay-cart with its mountainous load and the rakers and mowers in white shirts. The bittern hardly knows where to lay its eggs. . . . If the meadows were untouched, I should no doubt see many more of . . . the beautiful purple orchis there, as I now see a few along the shaded brooks and meadow's edge.

In the mid-nineteenth century, botanists reported this beautiful orchid to be a common species in Concord, but it's interesting that Thoreau's journal entry ascribes its absence from meadowlands to overly aggressive hay mowing.

In our modern surveys, the smaller purple-fringed orchis was among the rarest of the rare. After three years of intense searching for uncommon plants in the most out-of-the-way places, Abe finally managed to find a single purple-fringed orchis plant in the middle of a dense swamp near Annursnac Hill, as he described many years later:

As we were trudging through the swamp, the purple-fringed orchid stuck out when we came upon it. As I recall, it was rather tall with striking purple flowers and was growing on a moss-covered rock. We knew it was unusual and beautiful when we saw it — we didn't know, though, that it would be the only individual we'd see in all our days and years of looking for flowers in Concord.

With fallen stumps, catbriar thickets, and pools of muddy water, this home of the purple-fringed orchis is a difficult place to cross once you leave the path. I was convinced that there must be far more of these plants in this location, because it seemed so unlikely that Abe would have found the only one. Probably he had stumbled across one outlying plant of a large population. Or perhaps the plants were widely scattered in the swamp, and we only had to look more carefully to find dozens or hundreds of other plants. Yet even though we hunted throughout the surrounding wetlands, we never managed to find even a single additional purple-fringed orchis. The next summer, the plant Abe had located was gone. We never saw it again, and we never saw another purple-fringed orchis plant anywhere else in Concord.

This pattern of "missing" plants — species that Thoreau and others like him had seen, but that we couldn't find — repeated itself over the course of our study. Often we were unable to locate many of the species noted by Thoreau, Alfred Hosmer — who botanized in Concord at the end of the nineteenth century — or later botanists who studied the Concord flora. Or sometimes we found the missing species but in far fewer numbers than earlier observations. The conclusion is unmistakable: Concord is losing species. We have observed this pattern of missing species across numerous well-known and beloved plant groups, such as orchids, lilies, and mints.

Finding the Missing Wildflowers

Thoreau himself recognized that Concord was changing even in his own time and that many of the species of the past were no longer present.

> *Primitive Nature is the most interesting to me. I take infinite pains to know all of the phenomena of the spring, for instance, thinking that I have the entire poem, and then to my chagrin, I hear that it is but an imperfect copy that I possess and have read, that my ancestors have torn out many of the first leaves and grandest passages and mutilated it in many places. I should not like to think that some demigod had come before me and picked out some of the best stars. I wish to know an entire heaven and an entire earth.* (*Journal*, March 23, 1856)

In this passage, Thoreau was talking about the bear, deer, lynx, wolf, beaver, and other large mammals that formerly occurred in Concord. Such species were eliminated by the conversion of the landscape to farming and by hunting. Certain species, such as deer, were hunted for food, while others, such as wolves and bears, were deliberately exterminated because of their perceived threat to livestock. However, he was aware that human activities were degrading the wildflowers and trees as well: "It is well known that the chestnut timber of this vicinity has rapidly disappeared within fifteen years, having been used for railroad sleepers, for rails, and for planks, so that there is danger that this part of our forest will become extinct" (*Journal*, October 17, 1860). Thoreau would prove right about the chestnut's extinction, although it came about for reasons other than the one he imagined. Logging did not eliminate chestnut trees from Massachusetts' forests; rather, their demise was a consequence of a different human activity — the accidental introduction of chestnut blight fungus from Asia via imported timber or live trees.

During the course of our flowering-time studies, we searched Concord intensively to find the first open flowers of as many species as we could find. The more kinds of flowers we could find, the more precisely we could measure the impact of climate change on the plants of Concord. As carefully as we could, we tried to repeat the observations of Thoreau and Hosmer.

However, as we compared our list of observations with theirs, it was obvious that we had not seen hundreds of the species that Hosmer and Thoreau had observed. When we realized that we were not finding so many of the missing species, we wondered whether we were doing something wrong. So we turned for help to the modern naturalists who frequented the woods, fields, and swamps of Concord, such as Ray Angelo, Mary Walker, Peter Alden, Susan Clark, and Bryan Windmiller, asking where to look for the elusive species. Then we went back to the meadows and woods, redoubling our efforts. When we followed their advice and suggestions about where to look, we were sometimes rewarded with one of our "phantom" plants. But in many cases, a species was no longer where it had previously been seen in past years and decades, and remained on our list of phantoms.

Vanishing Orchids and Missing Mints

Although many varieties of plants have disappeared from Concord, the orchids observed by Thoreau illustrate this loss in species more dramatically than any other group of species. Orchids are intensively studied by botanists, because of their great beauty and distinctive floral shapes, which often are elegant and colorful, like clusters of miniature butterflies.

Richard Eaton — a botanist who worked for decades in Concord in the 1940s through the 1970s and who published a flora of Concord in 1974 — listed twenty-one species of orchids as being native to Concord, with one species possibly being extirpated. We were unable to locate fourteen of these species despite intensive searching. Of those seven species that we could find, five have been so reduced in abundance that they only persist as a single population. Of all of Concord's orchids, the fringed orchids constitute the largest groups of species, recognized by their large lower petal forming a lip, sometimes with fringes, and with a nectar spur projecting backward. Yet we were only able to find two species of the eight fringed orchids that have been collected in Concord in the past. In addition to the small purple fringed orchid Abe found in a remote swamp, we discovered that the green fringed orchid, formerly common, now occurs only in the Old Calf Pasture, where annual late summer mowing keeps the field open and allows the species to persist. With its narrow green leaves and pale green flowers, this species is hard to find among the sedges and rushes. Only repeated visits to the meadow and careful searching for wildflowers allowed us to discover the exact spot where it occurs. The numbers vary from year to year, but probably fewer than fifty plants are growing there.

Another orchid species, the nodding lady's tresses orchid, occurs as a few individuals in the yard of a private residence in the Conantum section of Concord. Fortunately, the home owner does not apply fertilizer or herbicides to the lawn, which would encourage the growth of grass and quickly eliminate the orchids. The nodding lady's tresses orchid provides a subtle beauty with its small white flowers arranged in a tight spiral around the stem. We did not see the other two lady's tresses orchids — the slender lady's tresses and the gracilis, which Thoreau had observed and described on August 23, 1853.

Whorled pogonia is another species that has always been hard to find in Concord. This rare orchid is listed as an endangered species in Massachusetts, and it occurs in one small patch of woodland in Conantum, where it was probably first discovered by Thoreau's sister Sophia. At this location, it does not even flower in some years, perhaps due to too much shading from the dense tree canopy. If the tree canopy above its location is not thinned out, the species may soon die out. The exact location is a closely guarded secret among botanists and local residents, as they are worried that orchid collectors will sneak onto the site and steal the plants for their wildflower gardens.

Finally, the rose pogonia, which Eaton had noted as common in the 1960s, now occurs only at Heywood's Meadow, right on the border with Lincoln near Walden Pond. Heywood's Meadow is one of the few remaining places in Concord where beaver activity has prevented swamp trees from growing up and shading out the bog species, including rose pogonia. This truly enchanting species produces a solitary dark pink flower at the top of a foot-tall stalk. The lip of the flower is tinged with yellow and has a distinctive fringe of hairs. However, only the most dedicated naturalist will find the rose pogonia in the town of Concord, as it is necessary to wade across fifteen feet of open water to reach the floating bog where it can be found. How many modern Concordians will be willing to wade through waist-deep bog water and roll themselves onto the edge of a floating bog to commune with the rose pogonia? Thoreau reserved a special judgment for this species, which he mentioned as "numerous" and called by the name "adder's tongue arethusa," noting on June 21, 1852, "The adder's tongue arethusa smells exactly like a snake. How singular in nature, too, beauty and offensiveness should be so combined!"

While there is still a single population of the rose pogonia, we can no longer find the beautiful small arethusa orchid. On May 30, 1854, Thoreau found it in abundance at Hubbard's Close: "It is all color, a little hook of purple flame projecting from the meadow into the air. A superb flower." This experience is now gone for the modern naturalists of Concord.

Orchids are not alone in disappearing from Concord. Other groups of species have also declined dramatically. Many aquatic plants, in particular, are disappearing. For example, Eaton listed eight species of *Utricularia*, or bladderworts, as occurring in Concord. Botanists actively seek these species because of their colorful, often bright yellow flowers, which resemble tiny snapdragons. The stems have small bladders that suddenly open when touched; through this mechanism, the plant can vacuum up small aquatic animals and then close, trapping the creatures inside. The dissolved prey provides minerals that the bladderwort can't get from the nutrient-poor bog.

Thoreau had a particular attraction to the purple-flowered bladderwort. On August 5, 1858, in Fairhaven Bay, a local enlargement of the Sudbury River, he counted a dozen flowers within a square foot: "They tinge the (water-lily) pads with purple for more than a dozen rods [198 feet]." The water lily pads are still common in Fairhaven Bay, but invasive species such as water chestnut and Carolina fanwort are now the most abundant species. Purple bladderwort, along with the lynx and the wolf, can no longer be counted as an active Concord resident. On a recent canoe trip to the bay, I did not see a single purple bladderwort flower. Moreover, of eight species noted by Eaton in his flora of Concord, only one species, the aptly named common bladderwort, is still prevalent. Another species, the humped bladderwort, occurs as a single population in a small pond near West Concord center. The other six species we could not find.

Another notably absent group is the mountain mints, species that typically grow in fields and woods. These minty-smelling herbs have opposite leaves and produce a dense cluster of small white flowers with purple dots produced at the end of the stems. Thoreau noted four species, his favorite being *Pycnanthemum incanum*, which had "prettily purple-spotted flowers swarming with great wasps of different kinds, and bees." Six species were apparently present in Eaton's time. However, we failed to find even a single individual of any mountain mint species at any location in Concord in any year.

Other botanists working in Massachusetts have also noted the substantial loss of wildflower species, again with particularly severe losses in orchids, mints, and certain groups of aquatic plants. Lisa Standley found that over 40 percent of the wildflowers had been lost in Needham, a suburb closer to Boston and more developed than Concord. In a study by Robert Bertin, the loss of species in Worcester County in central Massachusetts was somewhat less than in Concord, due in part to its larger size, which included some extensive tracts of less-disturbed landscape. The loss of wildflowers was also lower at the Harvard Forest, a more remote rural landscape.

Species Loss and Climate Change in Concord

Thoreau realized how delicately the world was in balance, and how easily that balance could be upset. Yet during his own time, he imagined that the world was in the greatest danger of slipping back into another ice age, rather than transforming into an overheated greenhouse. He wrote in *Walden*:

> . . . *nor need we trouble ourselves to speculate how the human race may be at last destroyed. It would be easy to cut the threads [of life] any time with a little sharper blast from the north, . . . a little colder Friday, or a greater snow, would put a period to man's existence on the globe.* (276)

Thoreau wrote these words based on his direct experience with the climate of his time. A century and a half later, my research team's experience stood in marked contrast to Thoreau's fears. Where Thoreau's journals contain frequent observations about cold temperatures, our field notes and conversations were more likely to include references to excessive heat — even when made during the same months and weeks of the year in which Thoreau was shivering! My instructions to new students emphasize the need to bring drinking water to avoid dehydration, rather than the need for warm clothing. And the end of a long, hot day in the field often includes a stop at Bedford Farms for ice cream or a cool milk shake.

It's interesting that Thoreau's quote about a greater snow that would "put a period to man's existence" doesn't address how the cold episodes he imagines might affect the natural world. In all likelihood, he believed nature to be much less fragile than humankind. Despite his awareness that direct human activities — such as logging, hunting, and farming — were affecting species and driving some of them out of the fields and forests of the Concord he knew, it may have been incomprehensible to Thoreau that human activities could affect nature on such a universal level. If he could have apprehended it, he no doubt would have been appalled. In modern Concord, we see evidence of a very different reality — one in which humanity's choices and actions, and not the natural shifts of a living planet, threaten our existence as well as the welfare of much of the natural world.

But *is* the loss of species in Concord related to climate change? Any scientist will tell you that correlation isn't causation. Species are missing, and temperatures are climbing, but did one have anything to do with the other? That's a puzzle that can be fairly hard to untangle, because much of

New England, certainly including eastern Massachusetts, has undergone a dramatic transformation in the time since Thoreau lived at Walden. Land-use patterns across New England have changed over the past century and a half, as agriculture has declined as a major source of livelihood and cities have grown and sprawled, becoming ever-larger "greater metropolitan areas." Paradoxically, in some areas, changes in land use have meant an *increase* in forest cover and even the restoration of some wetland areas, but they have also meant the construction of suburbs, shopping centers, and office parks. In Concord a number of major highways near and through the town bring additional disturbance into the mix. These and a multitude of related factors — such as air pollution, acid rain, increasing deer populations, and the arrival of invasive species — could all contribute to the decline of species. However, I think a case can be made for climate change as a prominent factor causing species to disappear, both directly and indirectly.

It took me a while to come to that conclusion. We were more than three years into our Thoreau research project before Abe and I came upon solid evidence that climate change *was* behind the loss of species in Concord. Abe and I had just given two successive lectures about our work to the botanists at the Harvard University Herbaria, Abe about changing flowering times in Concord and myself about the loss of wildflower species from Concord.

After my lecture, Chuck Davis, a young and enthusiastic Harvard botany professor, came up to me and asked if Abe and I ever had considered an analysis that combined our observations of changing flowering times with our data on species abundance. He proposed helping us with such an analysis. It was not clear to Abe and me exactly what Chuck had in mind, and it sounded complicated. As we both had other appointments that day, we agreed to meet the following week to discuss Chuck's proposal.

When we later met, Chuck explained that he and his research students could help us look at our two Thoreau-inspired data sets of flowering time and species loss in a new way. With his two grad students, Brad Ruhfel and Charlie Willis, he proposed using a phylogenetic approach, a sophisticated analysis that considered the evolutionary relationships among species, to look at the relationship between flowering time and species loss. The prospective collaboration was complicated by the fact that Abe and I couldn't fathom much of what they were suggesting — their statistical analysis was so intricate that even two scientists like ourselves had trouble grasping it. This didn't bode well for communicating our findings about climate change in a way that the general public could understand. In the end, Abe and I decided the proposed work had the potential to offer us valuable insights, and we agreed to collaborate with Chuck, Brad, and Charlie.

Chuck remembered the start of our work together:

> I am always thinking about plant evolution and how past climate has shaped where plants live. When Abe and Richard gave their talks at Harvard, it appeared that this information on flowering times and changing plant abundance fit perfectly into our studies of plant evolution. After Richard agreed to work with us, Charlie, Brad, and I were very excited to have two such novel data sets to link together. It turns out that this blend of ecology and evolution was the perfect approach to investigate climate change in Concord, and the results couldn't have been clearer.

A few months later, Chuck and his students delivered a report on their analysis of the Concord data — and their findings left us dumbfounded. The plants that had shifted their flowering times the most from Thoreau's time to the present were the plants that had best maintained their abundance in Concord over the past 160 years. In contrast, the conservative plant species that were still flowering at more or less the same time now as in Thoreau's time were the species that tended to decline in Concord over this period. This result clearly demonstrates that *the ability of species to respond to a changing climate directly affects their ability to persist in Concord today.*

The ability to shift flowering time over 160 years is also linked to plant families, with groups such as orchids, lilies, and mints showing less flexibility in their flowering times and a greater tendency to decline than other families.

We can look at the results in another way: instead of considering changes over time, we can examine the ability of species to "track" temperature in their flowering times. Budding plants respond to warming temperatures, switching on genes that control biochemical reactions. These reactions speed up developmental processes in warm weather and slow them down or stop them altogether in cold weather. Species that are good at tracking temperature will flower earlier in warm years and later in cold years. In contrast, species that are poor trackers will flower at more or less the same time each year regardless of the temperature, as their biochemical systems do not speed up or slow down as much in response to a varying climate. This further analysis done by Chuck and his team showed that species that are good trackers of temperature are also surviving better in Concord than are species that are poor trackers.

These interrelated results demonstrate that a warming climate is already causing some species, and certain entire families, to decline in Concord, with many species being driven to local extinction. The effects of climate change *are already here in Concord, right now*, and are already determining which species are winners and which species are losers. These simple and startling results tell us that global warming — something many of us consider to be a problem of the distant future in a faraway place, perhaps something that will affect our grandchildren or even great-grandchildren — has already begun its advance right in Thoreau's Concord. Species that can deal with a warming climate are able to persist and expand in Concord, but the species that can't deal with climate change are declining in abundance and are heading toward local extinction.

Why Do Flexible Species Succeed?

Leaving Thoreau aside for just a minute, why should a species that is good at adjusting its flowering time be a winner under a changing climate? What does the ability to adjust flowering time have to do with expanding plant populations? If you go to most scientific conferences and universities and ask this question, botanical experts and ecologists will most likely answer that species that can adjust their flowering time stay in synchrony with their pollinators, such as bees, butterflies, and hummingbirds. They will argue that the reason that the bulbous buttercup and the highbush blueberry are successful in today's Concord is that they shift in their flowering times at the same rates as their pollinators, so their flowers are well visited, have lots of pollen on the stigma, and set abundant seeds that can form the next generation. These pollinators might be bees, wasps, butterflies, beetles, hummingbirds, or even flies. In contrast, plants like the whorled wood aster and Canada lily, which are poor at tracking temperature, are not able to stay in synchrony with their pollinators. In consequence, their flowers are not as likely to be visited, the amount of pollen on their stigmas is inadequate, and few seeds develop. Such plants are not able to establish the next generation, and their numbers will decline over time.

This pollinator explanation makes good sense hypothetically, but it is unlikely to be the actual explanation. The great majority of plant species in the New England flora are visited by several different insect species; there are very few examples of plants that are so specialized as to be visited by a single insect pollinator. So even if a plant does not shift at the same rate as its pollinators, the plant will just be visited by other species of insects that happen to be around at the same time.

Also, the fruit and seed production of plants in Massachusetts are not usually limited by insect pollination. The rate of insect visitation and the amount of pollen carried to the stigma of the flowers are sufficient for seed production in most plant species that have been studied. Plants are more often limited by the amount of resources that they can devote to fruit production. This can be demonstrated by adding extra pollen onto the stigma of flowers by hand. In most cases, the percentage of flowers going on to bear fruit and the number of seeds per fruit will not increase very much after hand pollination. So even if flowers are visited a bit less with a changing climate, pollination intensity will probably not change the amount of seed production for these wildflower species. (This situation for wild plants is different for large-scale cultivated crops, such as apple orchards, where fruit production is often limited by the number of honeybees).

Finally, despite the large amount of research that has been carried out on the plants of New England, no one has described an example in which an earlier or later flowering of plants resulted in a pattern of greater or lesser fruit set. So while the pollinator hypothesis to explain the success of good shifting and tracking plants sounds logical, it is not really supported by any evidence.

A far more likely explanation, in my view, is that species that are able to change their flowering time in response to a warming climate are also able to change their leafing-out time. In many species, the timing of flowering is strongly correlated with the first appearance of leaves, with both events happening earlier in warm years and later in cold years. The reason for this is that dormant buds often contain shoots with leaves as well as flowers. Even when flowers and leaves are on separate shoots, the mechanisms causing bud break often are the same: warming temperatures and lengthening days cue both blooming and leaf out. I believe the reason that species that are good at tracking and shifting their flowering time are more successful in Concord is that these species are also leafing out earlier in warm years. Early leaf out gives them a longer growing season in warm years and a chance to accumulate more nutrients for growth, survival, and seed production.

Plants that leaf out early can also outcompete other species that leaf out later. If the bulbous buttercup can respond to climatic variation better than other species that grow in wet fields, and the buttercups leaf out earlier than these other species in warm years, then the buttercups will be able to grow bigger and outcompete other species. And in colder years that still occur sometimes, the bulbous buttercup will flower and leaf out later, and so avoid the frosts that kill other species. A further advantage to such flexible species that can flower and leaf out earlier in warm years is that they can complete their life cycle before the heat of the summer dries out the ground and prevents them from further growth. To me, this advantage of a more flexible leafing out is a better explanation for the success of plants that can track temperature and shift their flowering times.

How Fast Is Concord Losing Species?

The records of past botanists and our own observations can give us an appreciation of the pace of extinctions. Are extinctions in Concord a thing of the past, or are species in Concord still going extinct? Doesn't all of the protected land in Concord mean that these local extinctions have been stopped?

Sadly, most of the species extinctions in Concord appear to have occurred in the last three to four decades since Richard Eaton's flora was published in 1974, which was after much of Concord was already protected. Of 743 species that Eaton listed as having been in Concord during the nineteenth and twentieth centuries, not including grasses and other plants with inconspicuous flowers, he listed only twenty species as presumably extirpated, or locally extinct, at the time of his flora. In contrast, we could not locate over two hundred of the species from Eaton's flora in our recent explorations. What seems to be happening is that two dozen or so species disappeared from Concord during the period from 1839 to 1974 when Thoreau, Eaton, and other botanists worked, but hundreds of Concord's plants have been lost in the four decades since Eaton worked in Concord. It is possible that Eaton included any species that he had personally seen in Concord during the many decades that he worked on his flora of Concord and that he did not list specific species that had gone extinct by the time his book was published. In any case, the impression you get reading his book is that not many species had gone extinct from the nineteenth century through the early 1970s, yet we were unable to find more than two hundred of these same species.

That isn't to say that it's certain that all of these species we couldn't find are actually extinct. Thoreau and Eaton carried out their botanical investigations over a longer period than we did. Thoreau, in particular, walked almost daily in Concord for much of his adult life. From 1851 to 1858, he was very focused on plant life and says that he walked for at least four hours a day. Perhaps these past naturalists knew the locations of plants better than we did and saw more plants than we did. Or perhaps certain plant species are only evident in certain years or decades, and we missed them due to our shorter study. Thoreau explicitly mentions this need for extensive field observations:

> It is wonderful what a variety of flowers may grow within the range of a walk, and how long some very conspicuous ones may escape the most diligent walker, if you do not chance to visit their localities the right week or fortnight, when their signs are out. (*Journal*, May 8, 1853)

For the majority of plants, their most conspicuous sign is their flowers, while for others it is their early leaves in spring or their distinctive autumn leaf coloration.

Ray Angelo Gets Mad

The limitations of our study were brought home to us forcefully by the indignant reaction we received from Ray Angelo, a great admirer of

Thoreau who had studied the plants of Concord in the 1970s and 1980s, when, in the fall of 2006, we sent him a draft report of our observations of Concord, which included a list of plants we could not find. His e-mail reply in response to our report was bristling with irritation. He just could not believe that we could presume to say anything about a place as large as Concord, a block of land five miles across, after only four years of field-work. After all, Thoreau worked there for over a decade, Hosmer bota-nized for twenty years, and Eaton studied Concord for many decades. And Ray and his friends had also roamed around Concord for more than ten years. In Ray's mind, who were we to start writing anything about the flora of Concord after just four short years? Ray even issued us a specific challenge: he told us that he would choose fifteen species from our list of missing species and that he would find at least eight of them.

In the spring of 2007, Ray Angelo set out to prove us wrong about species having been lost from Concord. And, in his opinion, he suc-ceeded. He located populations of eight of our "missing" species and sent us their localities. Along Monument Street, he found one plant of wild ginger and a few small plants of blue cohosh. But we found that these sites were overgrown gardens and that the wildflowers were almost certainly planted, so we did not count them. Along a shaded sidewalk across the street from the Concord Middle School was a single non-flowering shoot of the sand cherry, and nearby were twenty small plants of the wavy-leaved milkweed, only one of which was in flower. And of greatest inter-est of all, in a wet forest next to the Assabet River was a single flowering plant of the roseshell azalea, probably the very same plant that Thoreau had been so excited to find on May 31, 1853, and called pinxter flower. Tho-reau had used all of his powers of persuasion to get his friend Melvin to show him where it grew, calling it "a conspicuously beautiful flowering shrub, with the sweet fragrance of the common swamp-pink [or swamp azalea], but the flowers are larger, and in this case, a fine lively rosy pink." Ray mentioned other species that he had found, but we were only able to locate two of these.

So, with Ray's ten years of experience in Concord, he was able to show us only perhaps five additional species growing in the wild, each at a single location. And of these five species, three consisted of only non-flowering plants, and two had only a single plant in flower. In Ray's opinion, he had proved to us that our work was inadequate, since he had found some of our missing species. From our perspective, although we appreciated this new information, it confirmed to us our basic findings: if the man who is currently considered the most knowledgeable person when it comes to the modern flora of Concord could only find five of our hundreds of missing species, and these mostly isolated non-flowering plants existing as single plants or small populations, then the flora is really undergoing the changes that we observed.

While we were initially shocked by Ray's first agitated message and his subsequent statements criticizing our approach, we came to realize that he had a good point. Any project tackling an area as large and fragmented as Concord is necessarily going to be incomplete. The key is to acknowledge the limitation of what we have done. In our case, we have acquired information on what plants can be seen by a team of scientists working for five years to that point (and eleven years as of 2013) and having the benefit of asking many other naturalists and local residents where to look.

We have continued to search Concord for missing species every year since then, with Chuck Davis and various other botanists joining in for the past two years. It is possible that some of these missing species still persist in Concord, in localities that we do not know about, as buried seeds in the soil or as inconspicuous non-flowering plants. However, we believe most of these missing species are likely extinct in Concord, as botanists have not been able to give us modern locations for them. We would welcome attempts by other botanists and wildflower enthusiasts to find these missing plants, and some will certainly be found. We continue to try to search for the missing plants ourselves. And Ray has gradually accepted that the general patterns we have found of species loss in Concord are valid, even though he disagrees with many of the specific details. In a February 20, 2013, article in the *Boston Globe*, Ray both criticized us and agreed with us, stating, "We and others believe, for example, that a list of 'lost' or 'extinct' species compiled by Dr. Primack and his co-workers represents more a list of what they happened not to be able to find rather than a list of what is truly gone. This is not to say there has not been change in Concord's flora for a variety of reasons."

Did We Forget about Deer?

Ray Angelo wasn't the only person to raise objections about whether the declines in species we'd observed were real. After we'd done a full analysis of the data — comparing Thoreau's observations against Hosmer's and our own — our paper concluding that species were disappearing from Concord partially as a result of climate change was eventually published in the high-profile scientific journal *Proceedings of the National Academy of Sciences of the United States of America*.

In response to this paper, several wildlife ecologists from Massachusetts wrote a strongly worded letter to the journal saying that our explanation for the changes in the Concord flora was wrong, or at least incomplete. These ecologists argued that the changes in the abundance of wildflowers in Concord that we had observed were almost certainly due to increased grazing by white-tailed deer rather than climate change. They further suggested that our results were more a question of finding what we wanted to find, rather than uncovering what was really the scientific fact of the matter. For academics, these are fighting words! They were calling us stupid or intellectually dishonest — and perhaps both!

We had recognized deer as representing a major factor causing a change in the plant life of Concord, along with habitat destruction, forest growth, altered land use, and pollution, but it is true that we had not included deer in our statistical analysis. In Thoreau's time, deer had been completely hunted out of Concord both for their meat and the damage that they did to crops. In recent decades, deer have come back to Concord; their numbers have increased rapidly with expanding forest cover and the absence of both hunting and their natural predators, wolves. Anyone who tries to grow lettuce or spinach in a backyard garden in Concord or almost anywhere in eastern Massachusetts can attest to the effects of the rebound in the deer population! In Concord today, deer graze on plants in the forest understory and munch on plants in the river meadows, and, yes, deer do have a powerful impact on wildflower abundance. In many national parks across the United States and elsewhere in the world, wildflowers have been drastically reduced in abundance because of the voracious appetites of unrestricted deer populations. In Japan I have seen strange forests at Nikkō National Park where almost every tree trunk is covered in an orange plastic mesh to prevent starving deer from stripping the bark off of trees in winter. And there are many experiments in the United States and other countries showing that when sections of forest have been fenced off to exclude deer, wildflowers become far more abundant.

But can deer activities explain the changes in species abundance that we have seen in Concord better than climate change? Can deer explain why some species have increased in abundance and others have declined? To answer this question, we — actually mostly Charlie — searched the published literature and found three articles where deer preferences for different kinds of wildflowers were reported. This list of what plants deer preferred and avoided included about 20 percent of the plants for which we had data in Concord.

When we (which again means "mostly Charlie") ran our analysis over again with this new information, we expected that adding in deer preferences would give us additional insights. However, adding in the preference of deer for certain plant species did not help to explain the changing abundance of plants in Concord. The ability of plants to change their flowering time was still a major factor in explaining which species were increasing or decreasing in abundance. So even though grazing by deer might explain the overall decline of wildflowers in Concord or even the decline of particular plant species, deer did not represent an important factor causing changes in the relative abundance of the flora of Concord.

But we had another prime suspect.

Purple loosestrife with monarch butterfly at Minute Man National Historical Park; photo by Richard B. Primack and Abraham J. Miller-Rushing.

Lysimachia stricta, *upright loosestrife,*
now well out, by Hosmer's Pond and elsewhere, a rather
handsome flower or cylindrical raceme of flowers.

THOREAU, *JOURNAL*, JULY 6, 1852

6. The Strife in Loosestrife

🌿 THOREAU OFTEN NOTED the profusion of wildflowers growing in Concord's river meadows and bogs, such as the upright or
yellow loosestrife. This perennial wildflower is also known as swamp candles for its long stalk of bright yellow flowers growing at the top of the
plant. However, in many of these same habitats, the native wildflowers
have been outcompeted and excluded by non-native newcomers such as
the purple loosestrife, an escaped ornamental plant from Eurasia. Fifteen
years ago, while driving around Boston, I became enthralled with purple
loosestrife and the profusion of papery purple flowers it produced on
abundant stalks above the foliage. I decided to try growing some in our
garden. I knew its reputation as an aggressive invader of wetlands; still, I
thought this might be an advantage, making the plant easier to introduce
into our boggy garden soil. I knew that sterile, non-seeding cultivars of
purple loosestrife were available for planting, but I was both a bit too lazy
to find out where to buy them and also a little curious to find out what
would happen in our garden with wild purple loosestrife.

The nearest purple loosestrife stand was in an old cranberry bog not
far from our house in Chestnut Hill. When I was a teenager, this field was
filled with a diversity of meadow wildflowers, including pink-flowered
swamp milkweed, different kinds of knotweeds with sprays of pink flowers, yellow-flowered bur marigolds, and willow herbs. But now the field
was a solid mass of waist-high purple loosestrife, so dense that few other
plants could gain a roothold.

One morning in what I will call Year One, I drove over to the meadow
and dug up four large plants, each with a root ball the size of a basket-
ball. I took them home and planted them in the rich, wet black soil of

our backyard. Unlike many plants that look tired and faded after transplanting, these loosestrife plants looked great. They immediately began to thrive and kept up a profusion of bloom all summer. I was pleased at the new addition to our garden . . . at first.

Purple loosestrife is emblematic of one of the most significant changes in the plant assemblages of Concord since the time of Thoreau: the great increase in the percentage of non-native species. By non-native species, biologists mean species that do not naturally occur in eastern North America. Some of these non-natives, such as purple loosestrife and various shrub and vine species of honeysuckle, were introduced on purpose, brought over the last two hundred years as exotic imports to beautify American gardens. Others were accidentally introduced as seeds through agricultural imports. Some of these species — such as garlic mustard, black swallow-wort, and Asian bittersweet, as well as purple loosestrife — have become dominant plants in certain habitats. These species are classified as "invasive" because they spread into natural habitats and exclude native species.

One traditional explanation for the increased abundance of these invasive plants has been that they are better suited than native species to changes in the environment, such as more disturbances by human activity, because they grow faster and can produce more seeds. The invasive species may be able to grow faster because they can absorb more nitrogen and other nutrients added to the soil and water by air pollution and fertilizer run-off from agricultural fields and lawns. There is additional evidence that many of these invasive plants could be responding to higher levels of carbon dioxide, giving them an extra advantage. Lastly, the insects, fungi, and other natural control agents that limit a plant's population size within its native range may not be present in its new range where it is invasive. Abe and I wondered whether climate change might also explain the success of invasive plants in Concord during recent decades. Could they also be responding to higher temperatures?

Charlie and Chuck were one step ahead of us, carrying out a new analysis comparing the flowering-time responses of native species and invasive non-native species. These results clearly identified a crucial factor of flowering time in the success of the invasive species in Concord. The invasive species were far better than the native species at changing their flowering time to track spring temperatures, flowering earlier in warm years and later in cool years. While both groups were flowering earlier now than in the past, the invasive species had shifted their flowering times eleven days further than the native species over the past century.

This result shows clearly that the success of invasive plant species is at least partly due to their ability to adjust their flowering time (and presumably their leafing-out time) in response to a changing climate. This physiological flexibility gives these invasive species a great advantage and allows them to outcompete native species and increase greatly in abundance.

What is noteworthy about this result, though, is that 80 percent of the non-native species occurring in Concord are not classified as invasive. We analyzed their flowering times as well. Based on their flowering-time flexibility, some of these non-native species have the potential to become invasive. One such non-native species that has shifted its flowering time is the mayweed or stinking chamomile. The flowering heads look like short versions of daisies that are produced low to the ground in great profusion in mid-spring, a few weeks earlier than true daisies. Their feathery leaves have a strong aromatic smell, and the species is related to the chamomile plant that is used in herbal teas. Mayweed is now flowering twenty-three days earlier than it did in 1900, a huge shift in flowering dates. The fact that mayweed has shown such an ability to change its flowering time suggested to us that this species should be watched for its invasive potential.

We mentioned this in an article that we wrote about our results, even though in Concord we have only found this species growing in one wet farm field. A few months after we published the article, we received an e-mail from a colleague working in the Himalayas telling us that mayweed was indeed taking over mountain meadows and should be considered invasive. It's exciting for us as scientists to have the ability to predict which species are likely to become invasive — but also somewhat saddening that this ability has come to us when it's really too late to stop many such species from spreading.

Our results also showed the curious result that non-native species had larger flowers on average than native species. This is almost certainly due to the fact that quite a number of the non-native species are ornamental plants that have large flowers, such as yellow iris and multiflora rose, which have escaped into the surrounding habitats. People start growing these plants in their gardens, and then they find that they have seeded into nearby lands, sometimes spread by birds. As you may have gathered from the story that began this chapter, I have been guilty of giving a "boost" to such an invader myself.

Let us return now to the purple loosestrife in my backyard.

The following spring, which I'll call Year Two, the plants came up again even larger in size, with many flowering stalks, each bearing hundreds of vivid purple flowers. But in the summer of Year Three, I noticed that we had dozens of small purple loosestrife plants flowering at new locations throughout our yard, and my neighbor Milton's yard had several purple loosestrife plants flowering as well. Milton was very pleased with this new addition to his garden and asked me what these gorgeous plants were. I realized that we were on the edge of a local catastrophe: we either needed to aggressively remove these plants, or the whole neighborhood was going to be taken over by purple loosestrife. So I ripped out and killed all of the loosestrife plants in our yard and the ones in our neighbor's yard too (when he was away). But that did not end the problem. Every year since then, I've noticed many small purple loosestrife plants in our lawn, growing from seeds and root sprouts, just waiting for a chance to mature enough to start flowering again.

This little experiment shows the potential of this non-native species to become invasive. While loosestrife displays a tenacity that's unusual even for an invader, it's hardly the only exotic species that can spread quickly given the right circumstances — and flexibility in flowering time in response to a changing climate only gives it, and others like it, a bigger advantage.

The Net Loss of Species

A hardheaded realist might wonder what all of the fuss is about regarding extinctions in Concord. Every ecosystem gains and loses species over time. In fact, if a large number of other species hadn't gone extinct in the past, it's likely that our own species *Homo sapiens* would not have evolved. Species have evolved and gone extinct throughout the history of life on Earth. You might ask (as many non-scientists do), *What's the big deal?* The problem is that human activities have caused extinction rates to increase by more than one hundred times over natural levels, while the rate of new species evolving remains at its previous low rate. The net result is that the number of species present on Earth is declining, with fewer species to provide the food, wood products, and medicine we need to live and the essential ecosystem services of flood control, water purification, soil formation, oxygen production, and carbon dioxide absorption necessary for the functioning of human society.

Concord is a microcosm of a problem we can see in many other conservation areas throughout the world: they are losing existing native species more rapidly than they are gaining new ones. Over time, there are fewer wild species living in a given place and fewer species in the forests and fields to enjoy and inspire the next generation of naturalists. In addition, as stated above and worth repeating, this loss of genetic diversity has serious implications for the health of individual species, ecosystems, and humanity. The Concord flora has had a net loss of native species since Thoreau's time, with about 83 species gained and an estimated 243 species lost. This represents a net loss of about one species per year. To my way of thinking, Concord is losing species at a surprisingly fast pace, given that about one-third of its land area is protected and another one-third remains undeveloped.

Because many native wildflower species have been reduced to one or two populations, often numbering just a few individuals, local extinctions will likely continue in the decades to come. Yet the situation is not hopeless. Habitat management, primarily to prevent swamp forests from replacing river meadows and bogs, appears to have allowed many rare wildflowers of open habitats to persist and recover. The removal of trees to create meadows along the Concord River at the Minute Man National Historical Park and the annual mowing at the Old Calf Pasture to prevent the growth of non-native shrubs show what can be done. Careful management of existing sites will likely be the key to protecting the species diversity of the historically significant Concord flora.

The McGrath Farm

We can also use land management to create habitat for specialized species. In the early spring of 2004, I realized that one group of plants that we had overlooked in the previous year was wildflowers and weeds of cultivated fields. While much of Concord had been cultivated during Thoreau's time, there are few active commercial farms in Concord today. Of these few remaining farms, most are not suited for wildflowers because they are show farms, in which neat rows of plants are being grown as much to create a pleasing landscape as to make money for the farmers, who may earn more income from farm-stand sales of products made elsewhere. But Abe and I had noticed one farm that was more pleasing to the eye of a botanist. On Barretts Mill Road in northwest Concord, there is a large farm on a river meadow next to the Assabet River. This farm had all kinds of crops: strawberries, corn, rhubarb, currants, and asparagus — and, what was best from our perspective, a great abundance of weeds. Even from the road, it was clear that this farmer did not do much weeding, nor was he using weed-suppressing herbicides. Growing among, beside, and even overtopping the crop plants was a profusion of weedy growth.

The old colonial farmhouse and barn matched the appearance of the fields, with rotting timbers, faded white paint, and, most impressively, vines growing in and out of the broken windows. To complete the picture were two ancient broken-down pickup trucks half covered in a thicket next to the farmhouse. This was the anti-Concord farmhouse, the antithesis of the multimillion-dollar gentleman's farm. Across the road was an attractive new farm stand, but it appeared to be permanently closed. I learned that, despite appearances, this farmhouse was not abandoned but was the property of Patrick and Michael McGrath, two brothers who had been living there as bachelor farmers. I decided that I needed to meet the McGraths and include their farm in our study.

Perhaps the McGrath farm could help us to reconnect with the weedy plants so well described by Thoreau in his account of farming:

Consider the intimate and curious acquaintance one makes with various kinds of weeds, — it will bear some iteration in the account, for there was no little iteration in the labor, — disturbing their delicate organizations so ruthlessly, and making such invidious distinctions with his hoe, leveling whole ranks of one species, and sedulously cultivating another. That's Roman wormwood, — that's pigweed, — that's sorrel, — that's piper-grass, — have at him, chop him upward to the sun, don't let him have a fibre in the shade. (Walden, 175)

The McGrath property also presented a mystery. On opposite sides of Barretts Mill Road, hidden by low trees, were two beautiful new country houses. My first suspicion was that the McGraths had sold house lots from their farm to a developer, enriching themselves by growing houses rather than asparagus on their land.

I was to discover that my initial guess couldn't have been more wrong. Over the next few weeks, I left many messages on the answering machine of Patrick McGrath, the brother who farmed the fields, but he never called back. In asking around town about the McGrath brothers, I learned that the farm and farmhouse had originally belonged to Colonel James Barrett (1710–1779), who was one of the officers in charge of the minutemen. The town of Concord had partnered with Save Our Heritage, a private land trust funded by a wealthy Concord resident, to acquire the house and the land from the McGrath brothers as a historic preserve, with the long-term goal of adding it to the Minute Man National Historical Park. As part of the deal, the McGrath brothers each were able to retain possession of part of the land, on which each brother had built his own house. Over the past five years, the trust has been restoring the Barrett house to its original condition at the time of the Revolutionary War.

Though I was learning more about the McGrath property, I still did not have permission to walk around the farm and search for weeds, and Patrick McGrath was still not returning my phone messages. Finally, one day in April when I was driving past the farm, I saw a new truck parked behind the farm stand. I parked my car next to the truck and headed to the back of the farm stand, where I found a man — Patrick McGrath, I later learned — carrying around some pots of vegetable plants. He was a large bear of a man, wearing old blue overalls, with a round face, long blond hair, and a blond beard. I introduced myself as a Boston University professor, and I explained that I was trying to make a list of the plants of Concord and wanted to walk around his fields to check what was growing there. He regarded me very skeptically and told me right away that this was not possible. "If I let you walk around, how do I know that you won't walk all over my strawberry plants?" he said. "And if you step in a hole and break a leg, how do I know that you won't sue me?" I explained that I was a botanist, and I knew what strawberry plants looked like; I promised not to walk on them. And I told him that I would not break a leg and sue him.

Despite these responses, McGrath remained unconvinced. He was quite clear: I could not walk on his land. Then I mentioned that it had been more than thirty years since Richard Eaton had done the last flora of Concord, and this farm was really important in knowing what plants were growing in Concord today.

"Eaton" was the magic word. McGrath suddenly became much more animated and responded that he had known Dick Eaton when he was a boy and had a copy of his flora of Concord. When I replied that I had met Eaton while he was working on the flora of Concord, suddenly the whole conversation changed. McGrath and I had established a connection. He now understood that I was someone who was continuing Eaton's work, and he gave me permission to roam around his farm, though he specified that I was not to walk near his house or his brother's. To this I readily agreed.

Over the years I have found many unusual plants on the McGrath farm, in many cases plants that I have not found elsewhere in Concord. The abundance of rare species at this one location is probably due to Patrick McGrath's organic farming methods, especially the absence of herbicides, and his particular tolerance for weeds.

One example of the treasures found in particular abundance at the farm is velvetleaf, a relative of hibiscus and cotton. This is a tall weed with large heart-shaped, hairy leaves along the stem and orange flowers at the top. Another species is the New England aster, a tall relative of the daisy with pale pink flowers. This species was listed as being common in the past in Concord, but it is now rare. Unfortunately, Patrick McGrath passed away in 2012, and the fate of his farm remains uncertain. Truly, Thoreau would have enjoyed meeting McGrath, who was probably the last farmer in Concord willing to provide equal space for crops and wild plants. Thoreau wrote of his own garden plot at Walden in terms that are just as apt for the McGrath farm today:

> [My field] was, as it were, the connecting link between wild and cultivated fields; as some states are civilized and others half-civilized, and others savage or barbarous, so my field was, though not in a bad sense, a half-cultivated field. (Walden, 171)

Any New Species?

In addition to the species lost from Concord, we observed eighty-three species that Eaton did not record as occurring in the wild in Concord.

Hosmer and Thoreau noted several of these species, which were probably grown in cultivation at that time but have since become established on their own. Escaped ornamental plants, such as longflower tobacco and garden lupine, represent the largest group of new species. In some cases, these have spread widely around Concord from seeds and cuttings, or they may be spreading just around where they were planted years or decades previously. Many new weed species and native species not previously reported from Concord also appear on the list of new species. The most abundant location for all such new species is along roadsides, parking lots, cultivated fields, railroad tracks, abandoned industrial sites, and dumping grounds. There are certain places in particular that we always check for new species, such as the West Concord train station and the Concord landfill.

In this list of new species, we have only included plants that are growing or spreading on their own. Even though it is exciting to find these new additions to the flora, their numbers are far fewer than the overall loss of species from Concord.

There are many additional cultivated species that were planted in specific places by people and are persisting in abandoned gardens behind houses or even after the houses and people are long gone. We did not count these persisting species, such as lilacs, as they are not expanding and reproducing new generations and are not really part of the flora. Thoreau felt deep emotion when encountering these last remaining living inhabitants of a forgotten family, as he writes in *Walden*:

> *Still grows the vivacious lilac a generation after the door and lintel and sill are gone, unfolding its sweet-scented flowers each spring, to be plucked by the musing traveller. . . . Little did the dusky children think that the puny slip which they stuck in the ground would root itself so, and outlive them and tell their story faintly to the lone wanderer half a century after they had grown up and died. (286)*

Minot Pratt's Introductions

Now that we know that so many native species have been lost from Concord, is there anything we can do to reestablish these species in the wild places of Concord? One of Concord's past residents carried out a lifelong project that bears directly on this question.

Minot Pratt (1805–1878) was a farmer and gardener who lived on Monument Street just north of today's Minute Man National Historical Park. Apparently as a hobby, he attempted to increase the number of species growing in Concord. Between 1860 and 1875, Pratt tried to introduce plant species to Concord that did not occur there on their own. He obtained plants from elsewhere in the United States, from climates that were similar to Concord's, extending from Massachusetts to Vermont, and even as far away as Illinois. Using his skills as a horticulturist, he attempted to establish these species at particular sites that were comparable to their native locations. For each species, he introduced numerous individuals at various locations in Concord, tending them for years to help their establishment. These species were often planted so skillfully that some later botanists believed these plants to be native to Concord.

Of the sixty wildflower species that Pratt attempted to establish in Concord (not including mosses, ferns, and gymnosperms), Hosmer noted that only seventeen species, or fewer than 20 percent, were still present in Concord in 1899, twenty-four years later. However, during our study, we were able to locate individuals of only a very few of these species in the wild. For example, we found several plants of the shrubby yellow root and the white-flowered false rue-anemone in the Estabrook Woods, near where Pratt lived. Pratt almost certainly planted these same individuals, and they have simply persisted, without spreading.

Certain of Pratt's plantings in the Punkatasset woods near his farm may have been eliminated by the expansion of horse fields in the 1980s by the current owners of the property. Other species that are still present, such as lily-of-the-valley and the large-flowered trillium, are common ornamentals that other people have certainly planted many times since Pratt's original plantings. If we assume that just two of the species that Pratt planted are still present as a result of his efforts, then the success rate is quite low, around 3 percent. In addition, of the seven orchid species that Pratt introduced, only one was still present in 1899, and none can be found alive today.

Considering the extent of Pratt's efforts, the results are quite meager and indicate that new populations of wildflowers are very difficult to establish — at least in a landscape undergoing as much transformation as has occurred in Concord. The take-home message is that we had better protect existing populations of rare and endangered wildflower species, because it is so difficult to reestablish them once they are lost, particularly when their habitat has been altered to suit other purposes.

There is one truly great botanical find close to the old Pratt farm: in the nearby woods are hundreds of saplings and dozens of moderate-size trees of the big-leaf magnolia, with its huge spreading white flowers produced at the end of its thickened twigs. This southern species is sometimes encountered as a garden tree, but to have hundreds of plants growing in a pure stand so far from any building is incredible. Not only did Pratt almost certainly plant them over a century ago, but they have obviously survived, thrived, and spread. Each year I try to visit this grove of magnolias at least once while they are in bloom, as a botanical wonder of Concord.

Thinking about all of our fieldwork in Concord and all of our data analysis, the final message comes down to this — if we look at the botany of Concord past and present, native and non-native, we find that global warming has changed Concord's flora in two distinct ways: First, numerous plant species are leafing out and flowering weeks earlier than they used to. And, second, plant species and families that can adjust their flowering times to warming trends are becoming more abundant; species and families that can't adjust to warming temperatures are disappearing.

And if plant communities are being altered in this manner, it seems impossible that bird, mammal, and insect communities aren't also being affected too. A warming climate will act directly on animals, changing their patterns of behavior and water needs, or act on them indirectly through their interactions with the plants on which they depend for food and shelter.

Taking Action

Thoreau was not just an observer of nature. He recognized the destructive tendencies of his fellow Concord residents and made recommendations on how to protect their natural environment. In one passage, he suggests the need for the town of Concord to establish conservation lands:

> *What are the natural features which make a township handsome? A river with its waterfalls and meadows, a lake, a hill, a cliff or individual rocks, a forest, and ancient trees standing singly. Such things are beautiful; they have a high use which dollars and cents never represent. If the inhabitants of a town were wise, they would seek to preserve these things, though at considerable expense.* (*Journal*, January 3, 1861)

Many studies, including our own, have shown that wildflowers and birds are declining in Massachusetts and other areas of the United States. Sometimes ecologists have been able to determine the causes of these losses and to suggest management strategies to stabilize populations and even help species to recover. In the 1960s and 1970s, when hawk and eagle populations had been declining due to the use of DDT and other pesticides, concerned citizens and scientists were able to get these chemicals banned; in the years since the use of these chemicals ended, raptor populations have recovered. Many wildflowers in Concord have been lost when the sunny river meadows were taken over by swamp forests after farming was abandoned; cutting the trees and the yearly mowing of these areas has allowed wildflowers to return in a few places. In the case of migratory bird species, such as flycatchers and warblers, the decline in numbers may be due to the loss of forest habitat in their tropical wintering grounds; international action is needed to deal with this threat. As we learn more about the threats to species, we can formulate actions to deal with them. However, climate change poses an entirely new type of threat because it means that protecting existing locations where certain rare and endangered species live may no longer be an effective conservation strategy. As conditions change, these species will not be able to survive in their present locations, even if they are protected from other threats. The conditions may become too hot or dry. A new strategy is needed to protect these species.

LET'S GET BACK to thinking about the rare wildflowers of Concord. The plants in Concord are presently adapted to live in a place in which the average April temperatures can range from around 41 to 45 degrees

Fahrenheit. This is based on what the temperatures have been in the past in Concord and the fact that the species have been growing in Concord for at least a century or more. However, temperatures in Concord have now increased to 43 to 48 degrees, with April 2010 reaching 50 degrees. Many of the wildflowers will not be able to tolerate these new higher temperatures and will gradually go extinct in Concord. They will still grow somewhere else in New England, just not in Concord. And temperatures in the Boston area and across the United States are predicted to continue warming during the coming decades as concentrations of atmospheric greenhouse gases continue to increase.

We've seen that a warming climate is a factor that is driving wildflowers to local extinction in Concord. There may be no point in protecting certain of these wildflowers in Concord, as the conditions are becoming too hot and too dry for them; they will suffer from water stress or other disruption to their basic physiology. If we want to protect these wildflower species, we need to take action to protect populations of them in cooler places farther north of Concord or at higher elevations, places like the Berkshire Mountains of western Massachusetts or areas of southern New Hampshire and Vermont. And if the species in decline in Concord are not at these places already, then we may need to bring the seeds and plants there to help get these species established. This process in which biologists help new species get established beyond their present range is termed *assisted colonization*. The new species that we introduce will replace other species that are declining and going extinct at those locations due to warming conditions.

This process of deliberately moving species in response to climate change makes sense to many biologists, myself included. We need to do it because we are changing the climate faster than most species can migrate by themselves. Of course, many birds, flying insects, and plants with wind-borne seeds can migrate quickly, but other species — such as salamanders, turtles, and plants with heavy seeds — will not be able migrate fast enough to track a changing climate and will just die out where they presently live. Also, the ability of species to migrate is hindered by the barriers created by human activity, such as highways, farms, industrial sites, and cities. The migration routes of such species will be blocked, and they will die out. So people will need to get involved in the process of assisted colonization.

A project involving assisted colonization will begin by carefully surveying a possible new site to determine if it has the right conditions of light, temperature, and water for the target species, both at the present time and for coming decades as the climate continues to change. The new site will have to be surveyed to make sure it does not have any of its own rare and endangered species that might be threatened by such a novel introduction.

Concord would be a good place to start the process of assisted colonization because the abundance and distribution of plants and animals there are already well known. The number of wildflower species in particular has been declining in Concord over time and will continue to decline as the climate changes. It is only by introducing new species from farther south that we can hope to maintain the number of plant species in Concord.

Concord does have many such places where we could plant warm-loving wildflowers from farther south, perhaps from southern Connecticut, New Jersey, or even Virginia and the Carolinas. These sites may also need to be managed in some way at the start, perhaps by removing or reducing some of the competing vegetation to provide places for the new species to become established.

Some biologists are strongly opposed to assisted colonization on the grounds that there is too great of a possibility that these species could become invasive. In my opinion, the greatest danger posed by assisted colonization is that attempts to establish new populations of rare and endangered species will fail, and the plants will die off at the new location due to competition with existing plants or grazing by local animals. The extremely low rate of success of Minot Pratt's wildflower introductions in the nineteenth century highlights the difficulty of establishing new populations of native plant species. The probability that the wildflowers will die off is far greater than the chance that the plant will become invasive. I cannot think of a single example where a rare or endangered plant species was moved a few hundred miles north of its present range within the same continent and suddenly became invasive. And I challenge anyone else to find such an example. We need to finish this endless academic debate and take action to save these declining plant species rather than simply document their decline and local extinction.

It is important to begin an active program of assisted colonization for a number of reasons. First of all, people created the problem of climate change for these species, and we owe it to these species to not drive them to local extinction. Second, all of the living creatures in the world are in some way interconnected, and if we damage the natural environment and the species that live there, we may ultimately create a world in which we cannot live. And, third, we also need to preserve nature for the next generation of people. Children growing up will want to see the whole world of flowers, butterflies, and birds, and they will ask us why we did not do a better job of protecting the environment. Lastly, we need to protect the diversity of species for the sake of a future Thoreau. Somewhere out there is, or will be, another genius who will examine the complexity of the natural world and give the people of the United States new insights into our society and perhaps ideas on how to solve our problems. I am always looking for a teenager or young adult who will be this new Thoreau. That is partially why I accept every invitation to speak about my work, particularly to groups of children, in the hopes that some will find their true calling to be a philosopher and a naturalist. There is one young adult I know who is always outside, making lists of birds and butterflies, taking close-up photos of tiny insects, and trying to make a living by writing about nature; but it is not certain yet if he will take what he knows and provide the kinds of insights that Thoreau did. The diversity of plant and animal species needs to be preserved in Concord and everywhere else for this future Thoreau, everyone else, and the species themselves.

Great blue heron in Walden Pond; photo by Tim Laman.

Birds are the truest heralds of the seasons,
since they appreciate a thousand delicate changes
in the atmosphere which is their own element,
of which men and the other animals cannot be aware.

L. JOHNSON, *THOREAU'S COMPLEX WEAVE*

7. The Message of the Birds

ONE EARLY SPRING MORNING in 1858, upon hearing the song of a purple finch, Thoreau noted in his journal, "How their note rings over the roofs of the village! You wonder that even the sleepers are not awakened by it to inquire who is there, and yet probably not another than myself observes their coming, and not half a dozen ever distinguished them in their lives."

While bird-watching is a fairly common pastime in modern America, in Thoreau's time interest in bird study was somewhat novel. Thoreau recorded with enthusiasm things that were happening around him in Concord, such as the songs of birds, that other people did not notice or regard as significant. Thoreau was also an early riser. In part this was a function of his times — most people of that era were up with or before the sun to take care of their animals and daily chores during daylight hours.

As a keen observer of nature in all seasons and weathers, Thoreau often recorded the first birds to arrive in spring. He also recognized that anything that altered the seasonality of plants would likewise affect other organisms — particularly birds and insects. To the modern biologist, his notes about seasonality sound especially insightful:

Vegetation starts when the earth's axis is sufficiently inclined; i.e. it follows the sun. Insects and the smaller animals (as well as many larger) follow vegetation . . . worms [caterpillars] come out of the trees; flycatchers follow the insects and worms; birds of prey the flycatchers; etc. Man follows all, and all follow the sun. The greater or less abundance of food determines migrations. If the buds are deceived and suffer from frost, then are the birds. (*Journal*, April 23, 1852)

Thoreau would probably be gratified to see how his fascination with bird behavior has expanded in his hometown. Concord is now a center of activity for bird-watchers, or "birders," boasting a large population of bird enthusiasts as well as a few famous bird experts. There's an annual Christmas bird count, in which well over one hundred people fan out across the town to count all of the species, trading stories of birds seen, birds heard, and birds that were observed just across the border in neighboring towns. Concord is also home to the recognized bird authority David Allen Sibley, the author of *The Sibley Guide to Birds*; originally published in 2000, it is arguably the best new field guide to North American birds written in recent years. Professors of ornithology from both Harvard University (Scott Edwards) and Tufts University (Michael Reed) and one of America's leading birding trip leaders (Peter Alden) also choose to reside in Concord.

One reason for this concentration of birders in Concord is the great variety of excellent bird habitat, including wetlands. The abundance of conservation land that made Concord ideal for my earlier wildflower work also makes Concord a haven for birds. A favorite place for birding in Concord is the Great Meadows National Wildlife Refuge, a 3,800-acre freshwater marsh in the northeastern section along the Concord River. Thoreau knew Great Meadows; in 1839 he and his older brother, John, traveled by canoe along the Concord and Merrimack Rivers. Although this wetland was not protected until 1944, it has remained a great birding location from Thoreau's time to the present — although its present inventory of species may differ substantially from what Thoreau noted there, as we will see.

The main features of the modern Great Meadows are two large shallow ponds beside the Concord River. The wetlands vistas are highlighted by the comings and goings of ducks and geese. There is a constant rustling of songbirds among the tall reed beds, including numerous red-winged blackbirds.

But no matter how early I set out, I'm never the first to arrive at the refuge. Large numbers of birders are already there, equipped with binoculars, spotting telescopes, and cameras fitted with telephoto lenses. They eagerly point out each unusual bird to one another and record their sightings on the blackboard in the kiosk on the edge of the parking lot. Birders check the board to see what species have been sighted that day and are encouraged to add their own new sightings at the end of their visit.

Birders have their own rituals. Most keep a list of species seen each day, birds seen over the season, and a "life list," a running list of species seen over that birder's lifetime. Many birders keep a personal journal and record when the first individual of a given species arrives in the spring (FOY, or first-of-year). Sometimes these journals are kept individually by a single birder, and sometimes they are kept collectively by members of a local birding club. Modern birders often enter their daily tallies into the Cornell Lab of Ornithology's eBird online database.

What I've learned through the years is that there are surprisingly large (at least to a botanist!) numbers of people who are interested in, even fanatically committed to, watching birds. Such people will get up at 4 a.m. to get to a site before dawn, when birds are noisiest and most active, making them easier to find by ear and eye. In the winter the most dedicated birders will stand for hours on a freezing, windblown beach on Plum Island to see sea ducks at a great distance with their spotting scopes. Other birders will travel to tropical mountains or the Antarctic to add species to their life list.

One April, during the second year of my Concord research, I visited Great Meadows to view the wood anemones, with their white flowers and dissected leaves that grow in profusion around the National Wildlife Refuge sign. On that particular day, while I was out looking for anemones among the birders, I was struck by a realization: just as the springtime blossoms might have a story to tell about climate change, so might the migratory birds that return from their wintering grounds when the weather gets warmer. This realization was to transform our Concord research project, taking it in a whole new direction.

Springtime Bird Arrivals

Over a billion birds representing 300 species migrate yearly into the continental United States, and over 150 species migrate through New England, some of them wintering as far away as Central and South America. These neotropical species migrate to take advantage of the great abundance of food, mostly insects, available in the spring and early summer to feed their young nestlings. Birds and their nestlings also probably face less chance of attack by temperate-zone predators, such as snakes and other arboreal carnivores, than they experience from the abundance of predators in tropical forests. Neotropical migrants fly south in the late summer or fall to avoid the harsh weather and lack of food of the long winter months. The early flush of insects in the spring and reduced number of predators must provide enough of an advantage to the adults to compensate for the large numbers of young birds that perish during their first year's migration.

The timing of bird arrivals in the spring is one of the most important potential sources of data on the impacts of climate change on biological communities. Migratory birds use a variety of cues to determine when to migrate in the spring, including temperature, wind direction, and day length.

The day length will let birds know the approximate time to begin migrating and will put them into the proper behavioral state of restlessness and activity, a condition known by the German word *Zugunruhe*, to prime them for migration. Day length, as perceived by the eyes and processed by the brain, is probably a major cue letting birds in the tropics know that it is time to migrate northward in the spring, particularly for birds that overwinter in Central America and the Caribbean. Near the equator of South America, where day length does not vary, birds may rely on other cues, such as the onset of the rainy and dry seasons, to time their migration.

For birds that overwinter in the southeastern and the mid-Atlantic United States, temperature may be the most important cue determining the timing of migration. If the temperature is warm, species will be more likely to migrate earlier, either because of direct action of warm temperatures on the birds themselves, resulting in hormonal or behavioral changes, or by causing the greening-up of the trees, shrubs, and herbs, and the emergence of early spring insects. In response to these cues, birds select days with dry, southerly air currents to begin their northward flight. Birds tend to avoid flying into headwinds, which require extra effort and deplete their energy reserves. They also will avoid flying in rainy weather, as the weight of water on their feathers requires extra effort and exhausts them, and also lessens their ability to maintain body temperature. All these factors mean that birds do not arrive in an even stream over the weeks of the migratory season, but rather in pulses associated with shifting wind directions and precipitation.

All of these many variables were on my mind as I considered the possibility of expanding my research program to include bird data. By accessing bird arrival times, perhaps I could determine if migratory birds are arriving earlier in the spring in the same way that the plants are flowering earlier in the year. Abe and I talked about these ideas, both between ourselves and with our various birding friends and ornithological colleagues.

We reasoned that birds should arrive as soon as possible on their spring breeding grounds in order to take advantage of the early appearances of insects available for adult birds to feed themselves and their nestlings. If a warming climate is shifting the greening-up of the plants and the emergence of insects to an earlier time, then birds must shift with them or risk losing a vital food source. Since most birds are territorial, an individual bird benefits by arriving earlier in the spring to stake out the territory with the most food, one that offers the greatest security against predators and other advantages. Birds arriving later might find that the best territories are already taken and be forced to raise their young on a marginal site. As a consequence, their nestlings might starve to death, die from heat exposure, or wind up as a snack for jays, snakes, or other predators with a taste for eggs and nestlings. On the other hand, if birds are too eager and arrive too early to the breeding site, they may not find enough food to eat and starve to death, or freezing temperatures may kill them outright. A migrating bird has to find the right balance between arriving early enough to get the best territory but not arriving too early. Timing is crucial.

Finding Bird Arrival Data

Even with this complexity of factors, analysis of bird arrival dates might provide evidence of the impact of climate change on bird migration patterns. Just as in our plant studies, we now needed to locate records of the first spring arrival dates of birds at locations throughout Massachusetts. Did those records exist, and if they did, how could we find them?

The answer proved surprisingly simple. Many of the most active birders of eastern Massachusetts are members of the Nuttall Ornithological

Club. Since I wanted to work with these birders and analyze their observations, I decided the best strategy was to join this club myself.

The Nuttall is a small ornithological society founded in 1873 by a Harvard professor, William Brewster, and a few dedicated birders. Through the Nuttall Club, I was able to meet the community of birders who record the first arrivals of birds across eastern Massachusetts, often at specific locations such as Mount Auburn Cemetery, a birding mecca on the border of Cambridge and Watertown. Many of these people work for universities, colleges, government wildlife agencies, and conservation organizations, such as the Massachusetts Audubon Society. Others are people with regular jobs but who share a passion for birds. I also renewed my friendship with Trevor Lloyd-Evans, a former Boston University grad student and now a professional ornithologist who has been mist-netting birds for decades at the Manomet Center for Conservation Sciences.

In 2004 I was voted in as a member in the Nuttall, and within a few months I was part of the network of Nuttall members. Through the club meetings, I learned that the birding community consists of people who love and study birds — birding isn't a casual hobby for most, but a way of seeing nature and a way of life. These people were willing to share their bird data with my students and me.

Peter Alden — Concord native, naturalist, and world-class birder — spoke about the role of birders as citizen scientists and the careful records they keep and share:

> Hundreds of birders use the Massachusetts Audubon checklist of birds to record the town and date of first observation for each of 300 or so birds seen each year. From that and more recent electronic trip lists, birders can find earliest arrival data quite easily. . . . Many birders are closet accountants and meticulously record all sightings and estimated numbers and comments at the end of each field day.

"There is so much data on birds," Alden said, "because birds are so visible, vocal, or both. It's possible to see between 30 and 130 different kinds of birds in a day. You would be hard-pressed to see a half-dozen mammals or reptiles in a day unless you were an expert tracker. Spotting that many fish would require an investment in gear or nets. Yet a pair of binoculars is all the gear required to study birds any day of the year."

Thoreau's Birds

In retrospect, it is perhaps not surprising that in my quest for bird data, of the various data sets that we found through my contacts in the Nuttall, the oldest records in Massachusetts came from Concord and were gathered by none other than Henry David Thoreau. Despite working on plants in Concord from 2003 to 2006, and despite our many contacts and friends within the community of birders, on the one hand, and with Thoreau scholars, on the other, *no one* had told us that Thoreau had kept a detailed record of bird arrival times in Concord! We found out only later that many of them knew about Thoreau's bird records. I only learned about the records because of a brief reference to them in a book that I was reading. When I asked Jeff Cramer, a Thoreau scholar and curator of collections at the Thoreau Institute in Concord, where I could find a copy of this list, he replied that it was at the Ernst Mayr Library at Harvard University — and did I really not know about this?

In the winter of 2008, I asked my new graduate student Libby Ellwood to go over to Harvard to see what she could find out about Thoreau's bird manuscripts. She recalled:

> I was eager to get across the Charles River to see what the holdings contained. These closely guarded gems could become a major chapter of my doctoral dissertation, so the stakes were high. I went over with an undergraduate assistant and was pleasantly surprised to find that his records were extensive, detailed, and in a form similar to a modern-day electronic spreadsheet. At first glance they appeared to be exactly what I needed. We wrote down as many notes as we could about the contents of these spreadsheets — species names, arrival dates, and locations. It was an incredible feeling to have Thoreau's original writings in my hands. At the same time, it was more than a bit nerve-racking to be leafing through these invaluable documents. I was happy to accept the librarian's offer to make copies of the spreadsheets for us to work with in the future.

When Libby called from the library to say what they had found, I could hear the excitement in her voice. And what a find it was! Thoreau had made a table of bird arrival dates during the 1850s, in the same style as his plant lists. This discovery added a whole new dimension to our climate change work. While some ornithologists had used Thoreau's list to determine which birds had been in Concord in the past and what their approximate arrival dates were, as far as we could tell, no scientists had ever thought to use this amazing time capsule of bird data to examine the impacts of climate change.

Once we had a copy of Thoreau's list, we put the species and dates into a computer spreadsheet for ease in handling. Then we began to investigate whether there were other records of birds' spring arrival times for other years in Concord. To our delight, we uncovered an abundance of such records beginning with the collected records of Brewster, the founder of the Nuttall Club and the American Ornithologists' Union. A Cambridge native, William Brewster had a passion to study birds, even though he had poor eyesight, and he made his earliest observations of birds in the fields, woods, and swamps of Cambridge.

Brewster recognized that the human impact on the local environment was causing great harm to many bird populations. In *Birds of the Cambridge Region of Massachusetts*, published in 1906, he lamented about the unfortunate "reclaiming — or, as some of us prefer to characterize it, *destroying* [of] — the Charles River Marshes, these once primitive and beautiful salt meadows" (33).

Most notable at this time was the decline in the great passenger pigeon flocks that had formerly filled the skies. In 1906, to secure a specimen of a male bird for his collection, Brewster shot a bird, commenting, "It was the last Pigeon I have seen, or am likely to see, alive in the Cambridge region" (*Harvard Magazine*, November 2007).

Brewster was an avid outdoorsman who liked to hike and canoe. He bought a tract of old forest on Davis Hill in Concord and built a cabin on it, where he stayed, often hosting his bird-loving friends. Using the cabin as the base for his excursions, Brewster observed and recorded Concord birds in general and bird arrival times in particular. Libby Ellwood was able to use Brewster's journals, which are housed in the Ernest Mayr Library, to extract first arrival dates for 1886 and the years 1900–1919.

The next key figure to observe Concord birds was Ludlow Griscom (1890–1959), also curator of birds at Harvard, a president of both the

American Ornithologists' Union and the Nuttall Club, and author of *The Birds of Concord* (1949) and *The Birds of Massachusetts* (1955). Griscom was a pioneer in the use of field characters and improved models of prismatic binoculars for identifying living birds, an approach that strongly influenced modern guides by Roger Tory Peterson and others. Griscom argued that "one need not shoot a bird to know what it was." He was reacting to the common practice of birders, before binoculars came into widespread use, to shoot first and identify later, as described by Peterson in *All Things Reconsidered: My Birding Adventures*:

> When a veteran ornithologist of an older generation wished to add birds to his collection, he drove out on a lovely May day from New York City to Van Cortlandt Park [in the Bronx] and was perfectly free to shoot as many warblers in the morning as he could skin in the afternoon. (217)

Griscom taught his methods of field identification using binoculars to Roger Tory Peterson, who expanded the approach in writing his widely used field guides to the birds of North America.

Griscom was particularly known for his studies of the birds of Central America, New York City, and Cape Cod, but he also recorded first arrivals of birds in Concord between 1930 and 1931 and between 1933 and 1954. These records are now housed at the Peabody Essex Museum in Salem, Massachusetts. As with Thoreau's and Brewster's records, Griscom's observations from Concord remained under-appreciated and under-utilized for decades, within plain sight of generations of researchers.

Modern Records

We next searched for contemporary records of bird arrival times. We were looking in particular for a single individual who roamed around Concord in the spring like Thoreau, Brewster, and Griscom. We found lots of people who liked to watch birds in Concord, and lots of people who had recorded bird arrivals for one year or a few years, but then had stopped. We also found a few Concord birders who recorded the first arrival of birds to their immediate neighborhoods. But where was the modern Thoreau, with decades of dedicated observations of all of Concord?

Through our contacts at the Nuttall Club, we found that person: a lifelong Concord resident and retired schoolteacher named Rosita Corey. Corey had made observations of first arrivals of birds from 1956 to 1973 and 1988 to 2007, during her frequent walks around the fields, forests, and wetlands of Concord each spring. Corey is also locally renowned as the only person who has participated in each of the past fifty Christmas bird counts in Concord.

We met her at her home in West Concord, close to the McGrath farm, to explain the work we had done on the plants of Concord and hoped to continue with birds. Corey quickly agreed to let us copy and use any information from her field books that we thought could be compared to Thoreau's observations. When she led us into her dining room to see the records, we were stunned.

Piled on the table were stacks of Audubon Society checklists in which she recorded on specified dates what birds she had seen and heard for the first time that spring, starting in 1956. There was a gap of fourteen years from 1974 to 1987 when Corey did not record bird arrivals. But even with this gap in her records, the amount of data was astonishing.

Even though the Corey data set was great, there was one peculiarity to it that was a slight weakness. About once or twice each spring, Corey would make birding trips to the town of Harvard, fifteen miles to the west, where her sister lived, and to Mount Auburn Cemetery straddling Cambridge and neighboring Watertown. The problem was that if Corey first saw a bird species during these excursions to Harvard or Mount Auburn, she would not record it again when she first saw it in Concord. This meant that, in certain years, no first arrival dates from Concord were listed for many common bird species. Because we were focused on Concord birds, we could not include the dates from Harvard and Mount Auburn in our comparison with the dates of Thoreau, Brewster, and Griscom.

Even with that small gap in the Corey data, we now had four great data sets spanning more than a century and half in Concord, giving us, as far as we know, the oldest set of observations of bird arrival dates of any bird study in the United States. These four sets of observations would allow us to see if the spring arrival times of birds were changing over the past century and a half in Concord, with Thoreau representing the early years, Brewster and Griscom the middle period, and Corey the recent decades.

How Good Are the Data?

Having four large sets of bird observations from different time periods sounds great and might be a lot better than what we had for our plant research. But plants don't fly away. If you want to identify a plant, you can walk right up to it. You can even photograph it (or, in the days before cameras, pluck it, press it, and carry it away with you) to compare to specimens or drawings if you need assistance identifying it. That's simply not possible with birds most of the time. Many birders obtain their "sightings" by little more than a flash of recognition, perhaps bolstered by hearing an identifiable call or song.

So is it really valid to compare these four sets of data? We are making the huge assumption that all four of these individuals were equally good at detecting the first arrival of birds and employed the same system of recording birds. We know that Griscom was superb at field identification, and that Brewster had poor eyesight. If Griscom was better able to detect the first arriving bird than Brewster, this would tend to make Griscom's dates several days earlier than Brewster. And we really don't know how capable Thoreau was at identification. He may have confused different bird species with one another, especially the most easily confused early spring arrivals, the warblers. This might have led Thoreau to record arrival dates for the wrong species. For example, on May 7, 1852, he wrote, "There appear to be one or more little warblers in the woods this morning which are new to this season, about which I am in doubt, myrtle-birds among them."

Since Thoreau, Brewster, and Griscom did not describe their methods, we don't know how they were identifying birds. Were they using sight identification alone to detect first arrivals or a combination of sight and birdcalls? We might imagine that Brewster relied more on songs and calls than the other observers due to his poor eyesight. By 1856 Thoreau could recognize the birdsongs; his journal includes notes about the song of the seringo, a species that we now call the savannah sparrow.

We also didn't know how hard our data collectors were looking. How many days per week had they each gone out to look for birds, and how many hours a day did they spend looking? For the most part, we don't know. We do know that Thoreau spent half of most days, sometimes longer, hiking around Concord, whereas Corey was a working teacher and would have had less time during the week to make observations. Griscom and Brewster both worked at Harvard University, and their responsibilities would have competed with their ability to observe birds every day. Knowing how much effort went into the search is important; the more days per week and hours per day that someone can devote to observing birds, the more likely that person is to record the day that a bird first arrives in Concord.

The more casual observer of birds, a person whom we might call a weekend birder, will on average record the same species a few days later. These factors can systematically skew a data set and lead to apparent differences in first arrival dates between time periods, when the differences are a result of how the data was gathered. It is important for researchers such as myself and my colleagues to be aware of such considerations, but in the end it is better to use the rich data that is available, while remaining aware of its limitations.

While we had reservations about the data we'd discovered, we decided to go ahead and use the observations made by all four observers to examine the impact of climate change on the arrival times of migratory birds to Concord. We had data on the arrival times of more than fifty species, but we were unable to use most of these species in our study. Some species that migrated in the spring to Concord in Thoreau's time are now year-round residents, including such species as the field sparrow, fox sparrow, swamp sparrow, and purple finch. The combination of a slightly warmer winter, a hospitable suburban landscape, and winter bird feeders give these former migrants a better chance of survival if they spend the winter in Concord instead of flying farther south.

There were also many species for which we had data for just a few years, making comparisons across time impossible. Most notable in this category were species that had declined greatly in abundance; these species included some that were recorded by Thoreau but were either quite rare or no longer present in Concord at all. For example, the Eastern meadowlark used to be abundant in Concord's meadows and fields, but now it is a rare visitor, as its favored open habitats have been largely replaced by forests.

After careful thought and lots of discussion, Libby and I decided to use the observations of twenty-two passerine (perching) bird species that had been seen by both Thoreau and Corey; Brewster and Griscom had seen most of these species as well. Overall, there were fifteen species that had been observed by all four observers. Using this limited list of bird species, we were able to compare 4 years of Thoreau's observations with 21 years from Brewster, 24 years from Griscom, and 38 years for Corey. Since Corey's observations had a 15-year gap in the middle, we decided to split them into early and late observations. Over the next few months, Libby and several undergrads carefully assembled all of the bird information from the different observers into a single electronic file and began to analyze the data. We were all anxious to know if we could get any useful results. Our plant data had given us startlingly clear evidence of the impact of climate change. Our bird evidence was a lot less clear.

What Do the Bird Data Show?

Perhaps I should have expected the bird data to be equivocal. After all, plants live their lives rooted to one spot, subject to the environmental factors happening in that one location. Plants can't move to escape the warm temperatures and drought. Migratory birds, on the other hand, move from place to place in response to cues of their immediate surroundings. They may travel hundreds, even thousands, of miles over the course of a season's migration. So it was only natural that we should have a harder time observing changes in behavior due to just climate change when there are so many more variables affecting birds' behavior.

The data for individual species highlight the challenges we faced using birds as indicators of warming trends. As we expected, there were three species that now arrive earlier than they did in the past — the Baltimore oriole, the yellow warbler, and the yellow-rumped warbler. But the majority — fifteen of twenty-two species — do not appear to have changed their arrival dates at all. Stranger still, we found that four species — the indigo bunting, ovenbird, wood thrush, and bobolink — actually arrive *later* in the spring than they did in the past. We struggled to make sense of these puzzling findings.

It is clear why arriving early would confer an advantage in a warming world. But what would be the advantage of arriving later in the spring, when a bird's principal food source of insects is almost certainly emerging earlier in response to the warmer temperature?

If temperature and sunlight were the only factors influencing bird arrival times, this question might be a great source of puzzlement, but the fact is there's a fairly simple explanation. Most of the bird species that appear to be arriving later also have declining populations, and such populations will have a contracting range of arrivals dates, with both later first arrival dates and earlier last arrival dates. For example, if the bobolink was abundant in Thoreau's time, we might imagine that two hundred birds arrived in Concord over a three-week period of May 1–21, with an average date of arrival on May 11. Now imagine that in recent decades, the bobolink currently has only has forty individuals in Concord, and the range of arrival dates is now May 6–16; the first arrival date has now shifted five days later, even though the mean arrival date is still May 11.

We could further imagine that the mean arrival date has also advanced in recent years to May 9, but with the declining population size, the first arrival date is May 4, which is still three days later than in Thoreau's time. So even when a species is responding to warming conditions and arriving earlier on average, a declining population size might influence first arrival dates. This is the challenge and problem of working with first arrival dates.

Another way to look at the effects of warming temperatures on birds is to examine the relationship of arrival dates with spring temperature — that is, to see how well a particular spring's arrival date matches up with how warm or how cool that springtime temperature is.

One advantage that we have working in Concord is that we can access long-term temperature data from the Blue Hill Meteorological Observatory in the Blue Hills Reservation, a state park southwest of Boston in Milton, Massachusetts. The Blue Hill Meteorological Observatory has been gathering weather records for the past 128 years. The weather records at Blue Hill are highly correlated with the weather in Concord: on a warm day in Concord, we nearly always find that it's also a warm day at the observatory twenty miles away. So if a species arrives in Concord in May, then we can look at how temperatures in eastern Massachusetts for March and April affect its arrival dates in May, based on the hypothesis that birds will arrive earlier in warmer years and later in colder years.

In theory, warmer temperatures in March and April would signal to the birds that conditions were warm enough so that the birds could survive, but also indicate the abundance of insect and plant foods for the birds to eat and to feed to their nestlings. The importance of such temperature effects would depend on the relative importance of climate and day length in determining when birds leave their wintering grounds and migrate north to their breeding grounds.

It may seem curious that we are asking how the weather in Boston in the one or two months before birds arrive affect their arrival dates — especially because you have to wonder how birds in Georgia or Florida will know what the weather is like in Boston. It might seem more logical to use the weather data from Atlanta to estimate when birds overwintering in Georgia should arrive in Boston. But it turns out that the temperatures from weather stations up and down the eastern United States are correlated with one another. Weather patterns line up so broadly across eastern North America that using just one weather station gives most of the information needed to determine if it is a warm spring or a cold spring across the region. Having data from two, three, or any number of weather stations doesn't give that much better an estimate of spring temperatures on the East Coast of the United States than just using Boston temperatures. In short, a bird that is living in Georgia in March can tell by the range of temperatures it experiences in Georgia whether or not it is warmer or colder than usual farther north in Massachusetts. If Georgia's March temperatures are unseasonably cool, the bird will wait longer to migrate north; if they're unusually warm, the bird may migrate earlier. So if we look at temperature data at Blue Hill and see that a particular spring was especially warm or especially cool, we can justifiably expect that the birds migrating north from Georgia or Florida would be experiencing this same yearly variation in temperatures.

When we use statistical tests to analyze how bird arrival dates in the spring have been affected by temperature since the time of Thoreau, we find that on average, birds arrive in Concord about half a day earlier per 1 degree Fahrenheit increase in temperature. Because Boston has warmed by about 4 degrees over the past 160 years, we could predict that species would be arriving about two days earlier on average in Concord. But, as described earlier, when we look at individual species, we don't see this general pattern of earlier arrival. The arrival data for any one species is just too variable from year to year. Perhaps this pattern would be more apparent if we looked at average changes across species in arrival dates. Such an approach would smooth out the variable responses of particular species.

So we did just that. When we looked at all of the twenty-two bird species together, we found that Thoreau observed migratory birds (considered as a group, all together) arriving in Concord, on average, on May 3 — and this date showed no change over time. Brewster's observations averaged slightly earlier, on May 2, and Griscom's averaged May 3, the same as Thoreau's. For Corey's early and later observations, the early decades averaged May 4 and the more recent observations averaged May 3. Given the large amount of variation among years for each observer, we saw no evidence that birds in Concord were responding as a group to warming temperatures by arriving earlier.

In fact, given the different methods of the observers, the changes in the Concord landscape, and the changing abundance of many of the bird species, it was striking how little difference we found among the five data sets.

We had to wonder: Why aren't the birds telling us the same story as the plants?

Climate Change and Many Other Factors

Our statistical analyses tell us that the birds arrive earlier in warmer years. So why weren't we able to detect any changes in their arrival dates over time if the climate is getting warmer? There are many possible explanations. One very simple reason is that there is great variation in the year-to-year arrival dates caused by the variation in climate in Boston; the variation in temperature, rainfall, and wind from year to year affect the arrival times of birds by days and even weeks. Birds don't fly in rainy weather, and they don't fly into headwinds. This year-to-year variation tends to obscure gradual long-term patterns, and such patterns are further obscured by declining population sizes over time.

We also know that migratory birds respond to temperature in different ways. It makes sense that where we don't see any changes in the spring arrival times of migratory birds as a group, we may still see changes in the arrival times of individual species. And this is exactly what we found when we looked at the arrival dates of particular migratory bird species — although even there, we found some surprises.

Overall, seven species have an earlier arrival date with warmer springs. The most responsive is the yellow-rumped warbler, which comes almost one day earlier for each degree Fahrenheit increase in the spring. We rather expected findings like this one; what we didn't expect was to find that some species, such as the wood thrush, actually arrive *later* with warmer temperatures, and that finding holds true even when we take into account declines in population size. For each degree Fahrenheit increase, the wood thrush arrives about 0.7 days later. This means that it comes earlier in cold years and later in warm years. Now, the yellow-rumped warbler's strategy makes sense — it arrives earlier in warm years so that its presence is most likely timed to the peak availability of the insects it eats. The wood thrush, on the other hand, does something a little odd; it tends to arrive *before* the peak of food abundance in cold years, and *after* the peak of abundance in warm years.

The later arrival of the wood thrush in warmer years is contrary to the prevailing view of why birds migrate, which is to take advantage of the abundance of food in early spring. So there are three possibilities. The first is that this apparent later arrival date is a product of bad data (that is, the observers simply missed seeing these birds until they'd been in residence for quite some time already). The second is a random event in which, just by chance, birds arrived late in some warm years and early in some cold years, leading to a false statistical relationship that has no biological basis. If this were true and we continued observing wood thrushes for several more decades, the relationship would disappear. The third possibility is that wood thrushes know something that biologists and other bird species don't know and are very much worthy of further study. Perhaps by shifting its arrival pattern in this way, the wood thrush can use resources in a way that other birds do not, and so in some way escape competing with them for food and territory.

We also found a geographic pattern to the bird arrival times. Birds that migrate relatively short distances, such as the Eastern phoebe and the pine warbler, arrive earlier in the spring from their wintering grounds in Florida and Central America. In contrast, later-arriving species, such as the red-eyed vireo and the Eastern wood-pewee, migrate all the way from South America, with the first arrivals probably coming from Colombia. The earlier-arriving species are more responsive to temperature than later-arriving species, which probably respond more to day length and changes in seasonal rainfall patterns. Also, there is a big difference in variability, with earlier-arriving species showing more variation in arrival times than later species. This makes sense, as the earlier-arriving species with short migrations are responding to temperature, which is highly variable from year to year, in contrast to the long-distance migrants that begin their northward journey in response to changes in day length, which are stable from year to year.

All of these results follow from the original discovery of Thoreau's list of bird arrival dates that had been kept for so many decades by dedicated librarians. Once we had that list, we were extremely lucky to find additional lists of birds from later time periods. I say "lucky" as there will be very few places in the United States (or anywhere in the world) that would have as many highly skilled ornithologists making careful, accurate observations over such a long time frame. What is even more remarkable is that these observations were preserved both in libraries and in the memories of living scholars and birders, who could direct us to where they were kept. Today in Concord, there are still hundreds of dedicated naturalists who keep up this tradition of recording bird observations, first started by Thoreau. This fills me with hope. Concord was a great place to start our work on bird migration, but it turns out that there were far more bird observations waiting to be discovered and analyzed.

*Cardinal with a data card at the Manomet Center for
Conservation Sciences; photo by Noah Reid.*

I found myself suddenly neighbor to the birds;
not by having imprisoned one,
but having caged myself near them.

THOREAU, *WALDEN*

8. Birds in the Mist (Net)

OUR SETS OF CONCORD BIRD DATA had major scientific value. The observations of Thoreau and others from Concord represented bird arrivals over a long time span, and Thoreau's enduring fame brought considerable popular attention to the project. However, one great disadvantage of the project was having only data of first arrivals in the spring. As we've seen, first arrival dates are affected when bird numbers change over time, and we know already that many birds in eastern North America have been declining over recent decades as suitable habitat is fragmented and reduced. A secondary problem was that, in the end, we really do not know if Thoreau and our other three observers were gathering data in the same way. To deal with this issue, Abe and I realized we needed to find a complete record of bird arrivals for a species that included the date that *every* bird arrived. Using this data we hoped to calculate the average date of bird arrivals each year, a value that was not affected by the number of birds. But where could we find such perfect data? Would anyone have this data, and would they be willing to share it with us? It turns out that the information was right in front of us.

One evening at a meeting of the Nuttall Club, while asking people about historical data sets, I bumped into my old friend Trevor Lloyd-Evans. Trevor is an Englishman who had been recruited to come over to the United States in the late 1960s to help establish the first comprehensive bird-banding station in the United States, the Manomet Bird Observatory, which would evolve into the Manomet Center for Conservation Sciences. Manomet is located near Plymouth, not far from the Massachusetts shore where the *Mayflower* first anchored in 1620. The center's original goal was to track where passerine birds were going on the eastern flyway as they passed through coastal New England on the way northward to their summer breeding grounds and southward to their wintering grounds.

Since 1970 Manomet has used the same procedure: catch birds in mist nets; record their species, age, and sex; attach a metal or plastic leg band; and release the bird. The Manomet data involves having living birds in the hand, rather than handling dead birds, which was common in Thoreau's time. The data from Manomet can be used to determine not only the date of first arrival of migratory species, but also the changing cohort size over time and the crucial *average* date of arrival for a species.

Trevor has been in charge of the banding project since its inception, working with a changing cast of interns and volunteers. This ensures that the methods at the station have remained constant over time. Trevor has an astonishingly detailed knowledge of birds. Confidently holding a live warbler in his hand, he distinguishes various species from each other using dozens of subtle differences in characteristics of color, wing, bill, and shape. He can even look at an individual bird and estimate how old it is, its overall health, and its likelihood of reproducing that year.

Trevor is one of those amazing people who, while strolling along a trail and listening to little chirps, tweets, and squeaks in the background, can name all of the birds in the vicinity, tell you how many individuals of each species there are, and tell whether they are male or female, juvenile or adult. And from the tone and frequency of the call, he will tell a story: "That noise we just heard was a male blue jay chasing a female Baltimore oriole out of its territory, right?" He will do all of this without raising his binoculars for a look and with an offhand style that suggests that he is just pointing out what everyone else knew and heard. He does it as if he is the secretary recording notes at a meeting, while all I had heard was some faint notes, barely distinguishable from the sound of wind or the rustling of the leaves.

What the Manomet Data Showed

The Manomet Center is about one hour south of Boston. Once off the highway, the secondary road passes the Plimouth Plantation, where "colonists" in period costumes reenact Pilgrim life in a re-created seventeenth-century village. A few miles past the plantation, there is a turnoff down a winding dirt road, passing by picture-perfect New England salt-box houses with gray shingle siding and through forests of oaks with an undergrowth of blueberry and huckleberry bushes. The center is perched on a cliff one hundred feet above the ocean, with a sweeping view of the Massachusetts Bay and the Cape Cod peninsula. Seals and cormorants can be seen sunning themselves on the rocks below. The center's low building houses a few offices and the bird-banding lab.

Manomet's researchers started catching and banding birds in 1966, and by 1970 they had developed a standard protocol that they have used ever since. The procedures developed at Manomet are simple and elegant. Along a network of straight trails cut through the low forest, fifty mist nets are opened each morning from half an hour before dawn through the day until half an hour after dusk. Each net is about thirty-six feet long and about eight feet high, with fine threads sewn in a mesh with square openings one and a half inches on a side. As they look beyond the nets at vegetation and insects, birds often cannot see the mesh, and they fly into the net and become entangled. Researchers check the nets once an hour, removing the birds carefully to prevent damage to delicate bones and feathers. Anyone working on the project must complete a certificate course in bird-banding techniques before they are allowed to remove birds from the nets. Such training is crucial, as birds sometimes further entangle themselves in their struggle to get free. Someone who is clumsy or in a hurry could easily harm or even kill a bird, breaking a wing or a leg, or removing so many feathers that the bird could not fly well. The nets also must be checked frequently, as an immobilized bird could freeze on cold days, overheat on sunny days, or attract the attention of a roving hawk.

Once a bird has been extracted from the net, it is placed in a small cloth bag and brought to the banding lab. There, it is removed from the bag, identified to species, measured for wing length, sexed, weighed, examined for body fat and sexual development, and aged. Once this information is recorded, a numbered aluminum band is attached to the bird's leg, with the band corresponding to the size of the bird — tiny tags for hummingbirds and kinglets, and large tags for woodpeckers. Once these procedures are complete, birds are returned to the wild, often just by placing them on the palm of someone's hand. (A staff prank is to tell visitors to take pictures of the birds as they fly away. The bird's takeoff is so fast and unpredictable, that no one is fast enough to catch the frightened bird on camera.)

Bird-banding at Manomet begins on April 15 and continues through June 15. Before April 15, there are too few migrating birds to make it worthwhile, and by June 15 the migration is mostly done. The nets aren't set out on rainy days, as birds don't migrate when it's raining, and even the birds that are present at Manomet don't fly or forage much on those days. On rare occasions, nets will be set out on weekends as a demonstration for children's groups or special visitors.

Over the last three decades, Manomet has netted more than 200,000 birds representing 159 species. That works out to be over 6,000 birds per spring season or around 150 birds per day of trapping. Most of these bird species were caught only rarely, too small a sample to provide usable data over time. Also, the total numbers of birds have been declining over time, and certain birds that were reasonably abundant in the 1970s are now rare. However, looking at the Manomet data, Abe and I found that thirty-two species were common enough that they could be used to determine changes in arrival dates over time; these species were reasonably common in both the beginning and the end of the thirty-two-year study period. Our study did not consider any of the few birds that were resident on the site year-round, a situation that developed during this period for a few formerly migratory species such as the blue jay and American robin. These resident birds could be recognized as they already had leg bands and showed evidence of being in breeding condition. Also, first-year fledglings were not considered, once they appeared on the site later in the spring.

How Do Changes in Population Size Affect Bird Data?

Before considering the effects of climate change on birds, we need to look at how bird populations are changing in size as they migrate through Manomet.

Birds in the eastern United States are declining for a variety of reasons. Among the reasons widely accepted by scientists and conservationists are the loss of their overwintering habitat as forests throughout the southern United States, the Caribbean, Central America, and northern South America are logged or cleared to make way for roads, power lines, farms, cattle ranches, dams, housing developments, and other human activities. The same habitat losses also affect their summer breeding grounds in the northeastern United States and Canada, as well as their stopovers along migration routes. When blocks of forest and other habitat become fragmented by roads, farms, and other human activities, harmful species such as rats, cats, raccoons, and parasitic cowbirds can feed on and destroy eggs and nestlings. As a consequence, many of these migratory birds species only successfully breed deep in forest zones — and such remote, undisturbed places are becoming harder and harder for the birds (or anyone else) to find.

Overall, for the thirty-two common species that Abe and I analyzed, the number of birds caught each year has declined by an average of one-third over the past three decades. Two species showed particularly dramatic declines: populations of the wood thrush and the Eastern towhee each declined by two-thirds or more since 1970.

For all of these species, we have information on when both the *first* bird arrives in the spring and when the *average* bird arrival time is. The Manomet data set is extremely unusual in having both of these types of data. Remember that the first arrival date is what most bird studies have and what we reported on for Concord. First arrival dates are affected by sample size, and we know already from Manomet that these birds are declining in population size over time. And we know that the average arrival time of birds in the spring is not affected by sample size.

The data from Manomet provide a clear message: the *average* arrival dates of eight of the thirty-two species are getting earlier over time. All of these eight species are arriving at least a few days earlier now than in the 1970s, and the Eastern towhee and the swamp sparrow are arriving one full week earlier than in the past. In contrast to the *average* date of arrival, none of these eight species has a *first* arrival date that is getting earlier over time. If we had looked only at the *first* date of arrival, we would have concluded that these species are not arriving any earlier over time.

Consider the Eastern towhee: in the 1970s, about seventy birds were caught per year with average arrival date of May 11, while in recent years only fifteen birds are caught per year with an average arrival date of around May 4. However, the first arrival of birds remains around April 23, with birds arriving earlier in some years and later in others. We've already seen that when populations decline in size over time, there are fewer early arrivals. This tendency of the first arrivals to get later over time due to declining population size cancels out the effects of earlier arrival times due to the impacts of warming temperatures. The net result is that first arrival dates can often show no changes in arrival over time when climate change is already having an effect. So when various studies using the first arrival dates of birds show either no effects or only slight effects of a warming climate on arrival dates, these studies are probably missing the real impacts of climate change.

The fact that climate change effects appear in many such studies suggests that a warming climate is probably having a profound impact on bird populations — and that this impact is almost certainly being underestimated.

The black-poll warbler provides another example of the effects of a declining population. As it travels northward to its breeding grounds in Canada, this warbler has a variable average arrival time in Manomet from mid-May to early June. But this species is not arriving any earlier over time, as judged by its average arrival time each year. In addition, during the past three decades, it has declined from around forty individuals caught per year to around twenty caught per year. As a result of this contracting number of individuals over time, the first arrival dates are also getting later. The first arrival of the black-poll warbler has now shifted by six days from about May 11 to May 17. If bird-watchers were monitoring this species for just first arrival dates, they would conclude that this species is arriving later over time and make a dramatic announcement about how it is responding in the opposite direction as other species. Similarly, this group might announce that the Eastern towhee is not showing any effects of climate change on its migration time based on first arrival dates; however, in fact, the average date of arrival is getting earlier.

This work from Manomet shows clearly the value of having the right type of data. Scientists must use first arrival dates when that is all they have available, but new studies to detect the impacts of climate change should always be set up to obtain average arrival dates.

While we were able to determine that declining population sizes can affect first arrival dates, we also realized that information on changing first arrival dates could potentially lead to predictions of the extent of population decline over time. If we look at how much each species at Manomet was declining and how many days later the first arrival was changing over time, we can see that for a "typical" bird species, a 50 percent decline in population size over thirty-five years can predict a three-day delay in arrival during this time period. We could then potentially apply this relationship to places where we have long-term observations of first arrival dates but no information on changes in population size.

For example, in South Korea, the first arrival of barn swallows has been recorded at government meteorological stations across the country for the past fifty years, but there are no long-term records of changes in swallow abundance. Many of these barn swallow populations are now arriving three to four weeks later than they did previously. Anecdotal evidence suggests that their populations are declining as flying insects that are their main food source become scarce, probably because of increased pesticide use and other changes in farming methods. By using the number of days that each Korean swallow population is arriving later in the spring, we can estimate the percentage decline in population size. Based on this interpretation, we have estimated that many swallow populations have declined by about 95 percent or more across South Korea over the past fifty years — although this prediction needs to be confirmed by field studies.

Birds and Climate: What Drives Arrival Times?

With this excellent data set from Manomet, we are able to identify the climatic factors that are actually driving the changes in average arrival dates of these thirty-two species. When examined by individual species, spring temperatures in March, April, and May are the best predictors of the spring arrival of ten species that are short-distance migrants from the mid-Atlantic states, with all species arriving earlier in warmer weather. For example, the hermit thrush arrives one day earlier at Manomet for each 1 degree Fahrenheit increase in temperature, and the swamp sparrow arrives 1.4 days earlier for each 1 degree increase in temperature.

In addition, six species, some of which also showed temperature effects, showed a linkage to the El Niño/Southern Oscillation (ENSO) Index, a broad climatic index linked to the temperature of the central Pacific and increasingly recognized as having climatic impacts throughout the world. Positive ENSO values cause dry, warm weather in the southeastern United States and cool, wet weather in Central America, the southern Caribbean, and northern South America. Negative ENSO values (a phenomenon known as La Niña) result in a reversal of the weather phenomena. Mid-distance migrants from the southern United States are most affected by the ENSO Index, which is also linked to wind direction.

Another large climate pattern is the North Atlantic Oscillation (NAO) Index, in which positive values correspond to more southerly winds, and negative index corresponds to greater precipitation. The hermit thrush and the American robin both arrive earlier in years with a positive NAO Index; that is, with more southerly winds.

When we consider all of these temperature data and the two climate indices, we are able to explain 13 to 30 percent of the variation in mean arrival times of sixteen of the thirty-two bird species at Manomet. That is, other factors that Abe and I have not considered explain a further 70 to 87 percent of the variation in arrival times. So, should Abe and I be happy that we can explain some of the variation in arrival dates? Or should we be depressed that with so much effort at catching birds and analyzing one of the best bird arrival data sets in North America, we are still a long way from explaining most of the variation? In fact, most scientists would be glad to explain this much of such a complex system. The various factors of wind direction, alterations in flight path, and daily rainfall might cause some of the variation — and let's not forget the variation introduced by not catching birds on Saturdays and Sundays and so missing two days in a row when birds might have come and gone without being detected. And of course, out of the thirty-two birds, there are also sixteen for which we were unable to explain any of the variation in arrival dates. The great crested flycatcher and the brown thrasher have arrival dates that none of our variables can explain. Our analysis sheds no light on what affects the year-to-year variation in arrival dates for those two species.

Abe's and my work with the Manomet data lasted five years, from 2003 to 2008. Libby Ellwood is currently analyzing the records from

Manomet to determine if bird species that are able to adjust their arrival times are better able to maintain their population sizes. Throughout our eleven-year association with this project, my students and I have headed down to Manomet every spring when the birds began migrating, largely for the purpose of keeping up our relationship with Trevor and to emphasize that we are still actively engaged in this complicated project.

Libby Ellwood recalled the visits this way:

> Traveling to Manomet each spring to visit Trevor and observe the happenings at the banding station is always a thrill. It is such a unique experience that I simultaneously want to keep it a special secret among just myself and the banders, and to share it with everyone I know. If everyone was able to experience walking the trails, searching the nets and holding the colorful birds in hand, I am certain the world would be a better place. This is, of course, at the cost of the welfare of the birds, and hence a desire to keep the entire operation under wraps. It is awe-inspiring, though, to consider that these tiny creatures are nearing the end of an arduous journey hundreds, if not thousands, of miles long to reach their northern breeding grounds. There could be no better spokesperson for migratory birds than Trevor, who knows every species and describes each as if describing a member of his own family.

A Farm Diary

Concurrent with our work in Concord and at Manomet, we tried to locate other useful data sets, especially the diaries of bird arrival dates kept by individuals. A personal journal kept by an individual has the advantages of a consistent set of techniques for observing and recording data.

Aside from Rosie Corey in Concord, the best diary that we have found is one kept by Betty Anderson, a dedicated naturalist who has lived on the hundred-acre Wolf Trap farm in Middleborough, Massachusetts, since 1950. Anderson is a committed conservationist who has been active in wildlife protection at the local, state, and national levels her whole life. From 1969 to 1983, she was the director of the fledgling Manomet Bird Observatory. Since the 1960s, Anderson has kept a diary of the birds and other wildlife that she has seen on her daily walks around the woods, swamps, ponds, and fields of her property. She made many of these observations while sitting at her kitchen table, where she had a clear view across her fields to a pond.

I first heard about Anderson in 2003 from Trevor Lloyd-Evans, who thought she might have made some systematic observations of birds on her farm. When I telephoned her, she was initially quite hesitant about whether her diaries had any scientific value. However, we agreed that it was at least worth a look. A few days later, I headed south with two under-grads, Anna Ledneva and Carolyn Imbres, to spend a day at the Anderson farm.

When we arrived, Anderson graciously welcomed us into her home, a comfortable farmhouse from a bygone era. On her living-room table, she had piled up some of her yearly diaries for us to look at. In these diaries she had recorded the dates of first arrival of birds in the spring, the time of first calling of various frog species, the first appearance of butterflies, and the first flowering of many common trees, shrubs, and wildflowers. Inter-mixed with these records of spring events were descriptions of vacations and field trips, and notes about visits with family and friends. After look-ing at the diaries, it was clear they had considerable scientific value. With Anderson's permission, we decided that we should make a list of plant and animal species that she had observed in most years and start entering dates into a spreadsheet. This was a substantial project that would require many visits to her farm.

A pattern began to emerge; once a week Anna, Carolyn, and some-times Abe and I would sit down in the Anderson living room. Betty An-derson would give us a large plate of homemade chocolate-chip cookies, and while munching away, we would carefully turn the pages of the dia-ries, recording the first appearance dates of the common species. There were sixteen bird species that were common enough through the years, many of which could be seen readily from her kitchen window, such as the ruby-throated hummingbird visiting her nectar feeders, barn swallows and tree swallows flying over the fields catching insects, red-winged black-birds perching on tree branches along the pond and displaying their scar-let shoulder patches, and wood ducks on the open pond. Anderson also noted the first flowering of three plant species: goldthread, notable for its small dark green, three-parted leaves and bright gold-colored roots; wood anemone, with delicate dissected leaves and white flowers; and spicebush, with its abundance of tiny yellow flowers and lemon-scented leaves. On her list were also the first calling times of the American toad, the spring peeper, and the wood frog; and the first appearance of two butterflies, the spring azure and the mourning cloak.

Earlier Appearance of Spring Activity

Of all of these twenty-four signs of spring — including sixteen bird species, three amphibians, three plants species, and two butterflies — five species, all of them birds, showed a tendency to arrive earlier over the past four decades. These were the wood duck, the ruby-throated hummingbird, the ovenbird, the chipping sparrow, and the house wren.

The wood duck showed the most dramatic change, arriving on average more than thirty days earlier now than in past decades. In the 1970s wood ducks generally arrived in early to late April, but in recent years they have been arriving in March and sometimes even late February. It is possible that the wood duck arrived even earlier but remained unobserved in the nearby forest, only being noticed by Betty Anderson when the ice melted on the pond and the ducks settled on the water. As Anderson herself pointed out, the record of first appearance of wood ducks is really an indirect measure of ice-out on the farm pond. Thoreau also noted in his diary on March 18, 1853, the rapidity with which ducks seek out the first open water in his early spring: "How eagerly the birds of passage penetrate the northern ice, watching for a crack by which to enter. Forthwith the swift ducks will be seen winging their way along the rivers and up the coasts. They watch the weather more sedulously [diligently] than the teamster [wagon-driver]."

Anderson would have certainly noted the wood duck, as this species is considered one of the most striking waterfowl in the world, with iridescent plumage, green head with a crest, red eyes, and bold white lines accenting the head, throat, and chest.

The ruby-throated hummingbird, the only hummingbird in New England, is also arriving two weeks earlier now than it did previously at Anderson's farm. Perhaps part of the explanation for this rests in the dramatic increase in the number of people who put out hummingbird feeders, providing a valuable supplement to the diet of these tiny, hyperactive birds when they arrive in the spring. The thrumming sound of their wings, their hovering flight while visiting flowers, and their brilliant green coloration make these one of the easiest birds to identify, and one of the most amusing and enjoyable to watch.

As we did with Concord birds and plants, we can also examine how Anderson's species respond to late winter and early spring temperatures. Of these twenty-four species, the wood duck, the red-winged blackbird, goldthread, and the spring peeper frog all show clear responses to temperature, with earlier activity in warmer years. The strongest response is again shown by the wood duck, arriving three days earlier for each 1 degree Fahrenheit increase in temperature.

The most important conclusion coming out of the analysis of Anderson's diaries is that one person keeping a diary can provide important information on the ecological response of species to climate change. Anderson took a few years to learn to recognize the species in her area, and thereafter made faithful observations of the birds, amphibians, plants, and butterflies around her property. She didn't make observations every day, and she sometimes left on vacations. Yet her diaries have scientific value because she was persistent: she was confident of her identifications, she kept records year after year, and she preserved her diaries in good condition. Perhaps most importantly, Anderson was willing to share her observations so that they could be properly analyzed using modern statistical techniques and then published. Anyone interested in nature or climate change could begin making these same types of observations around where they live, walk, or work.

Putting It All Together

With a variety of graduate students, such as Abe and Libby, various colleagues, such as Trevor Lloyd-Evans and Betty Anderson, and many Boston University undergraduates, I began this bird work in 2003 and have continued with it for over a decade. When we first started, there were no major studies on the effects of climate change on bird migration anywhere in the eastern United States, and only a few studies from elsewhere in North America and Europe. We were a bit shocked in 2003 when Chris Butler, an ornithologist of whom we'd never heard, published a study in the *Ibis*, a good bird journal, analyzing a century of records gathered by the Cayuga Bird Club in central New York and the records of the Worcester County Ornithological Society from Massachusetts. Butler's study indicated that there was a widespread pattern of birds arriving earlier in central Massachusetts.

When Butler's paper was published, at first Abe and I worried that we had been "scooped." As Abe recalled:

> Because the field of phenology and climate change was so new and moving so quickly, it was hard to know where our work stood relative to the other work taking place. New papers were coming out with some frequency. When Butler's paper came out, making some pretty interesting findings from a huge data set of bird migrations pretty close to where we were working, my heart certainly sank a bit. It made me question how new our work would be and whether we could get our results out in time to make them relevant. In the end, those concerns were unwarranted, but I had no idea at the time.

But Abe and I quickly concluded that there were many reasons why we should continue with our own research, most importantly because we wanted to find local data to compare with our studies of Concord's plants. So Abe, Libby, and I did continue our work on birds, and we produced a tidy flock of papers on the impacts of climate change on the birds of Massachusetts. Certainly the most important finding is that although some bird species are now arriving earlier in the spring than in the past, the rate of change for birds is far less than it is for plants.

Clearly a key next step is to make careful studies of the food sources for the newly arrived birds. These are the insect populations that emerge in the spring — the insects that feed on the plants and are in turn eaten by the birds, and later fed by parent birds to their nestlings. However, there are over a hundred migratory bird species, and each one eats something different and finds its food in a separate way. Brown thrashers turn over leaves on the forest floor, swallows glide over fields catching flying insects, warblers glean insects from young leaves and flowers, and flycatchers make short flights out from branches to snatch passing insects. Each of these birds is catching different insect types, and probably any one bird species may be feeding on different insect species across its range depending on what is available. Perhaps the flycatchers such as the Eastern phoebe will be eating mostly mosquitoes and gnats in one part of their range, large flies in another part, and beetles someplace else. So to determine how climate change is affecting a bird's food source, the first step is to determine what the birds are eating and then to determine how climate change is affecting those insect species. Such long-term studies should be started as soon as possible and are a major priority for climate change research. Could part of the reason that many bird species are declining be the lack of synchrony between the arrival of birds in the spring and peak insect abundance? Has climate change thrown this crucial ecological relationship out of balance? Or are the changes in bird populations due primarily to habitat destruction?

Another important new line of research is determining how milder weather in the autumn affects the timing of when birds leave their breeding grounds in New England to head south for the winter. There is some evidence that certain bird species in North America and Europe now arrive earlier in the spring and then depart earlier in the summer once they finish breeding. In contrast, other species are now taking advantage of the longer growing season to have additional broods, and so are both arriving both earlier in the spring and leaving later in the autumn. However, it remains to be seen if this pattern holds up.

Feathered Ghosts

We were able to carry out our research in Concord because of a tradition of local naturalists studying the birds, wildflowers, and other species of Concord, along with a strong conservation ethic. While new homes, office buildings, roads, soccer fields, and tennis courts fulfill important needs of human society, the homes and foraging grounds of Concord's birds also need to be preserved. Without the actual birds and associated Concord environment, much of the basis of Thoreau's great gift to the public will be lost, and the next generation of young naturalists will find the forests and fields to be only filled with the ghosts of birds.

In Concord today, there is great effort being made by government agencies and private foundations to protect the old buildings and sites associated with the Revolutionary War battles and the places where Thoreau, Emerson, and other Concord philosophers lived. We also need to preserve the places where the birds of Concord live today. Thoreau spoke forcefully of the need for such protection, remarking on September 6, 1850, upon seeing a great blue heron shot by a neighbor, "I am glad to recognize him for a native of America, — why not as an American citizen?"

And if the great blue heron and other birds are truly American citizens, then they have rights and should not be harmed, either when they are flying from their winter resting places to their spring homes farther north, or when they are resting in their Concord homes or at work gathering their food.

Some years ago I came home on a spring day to find that our new neighbors had cleared their wooded backyard to make a lawn. One of the trees cut down was an old red maple with many holes where downy woodpeckers gathered their food each day and had made their nests. The neighbors and the landscaping crew had surely not asked the woodpecker family if it was acceptable to destroy their home and probably kill their nestlings. Even if my neighbors really needed to make such a lawn, couldn't they have waited until the woodpeckers had finished raising their family and were ready to move on?

In addition to the rights that birds have all by themselves, some people recognize that birds provide valuable roles in the functioning of the ecosystems on which we depend. Birds provide a diversity of critical functions in the New England landscape, including pollinating flowers, dispersing seeds, and eating insects. Without ruby-throated hummingbirds, the beautiful electric red spikes of cardinal flowers would not be pollinated and set fruit, leading to the decline of this wondrous species, as the current generation of plants is not replaced. Many forest tree species rely to some extent on birds to feed on the insect populations that would otherwise expand without control and defoliate the trees. The list goes on and on. Without birds to disperse their seeds, numerous plant species would not be found in our forests, fields, and wetlands. And in turn these plant species provide the birds and us with wood, purify water, control floods, protect the soil, and even produce the oxygen that we breathe.

Thoreau was one of the first naturalists to understand the role that birds play in dispersing seeds and establishing new forests, writing:

I have often wondered how red cedars could have sprung up in some pastures which I knew to be miles distant from the nearest fruit-bearing cedar, but it now occurs to me that these and barberries, etc. may be planted by the crows and other birds. (*Journal*, February 4, 1856)

In fact, for numerous trees, shrubs, and wildflowers — such as black cherry, sassafras, high bush blueberry, huckleberry, and wintergreen — birds eat the fleshy fruits, and then either regurgitate the seeds later or pass the undamaged seeds undigested in their droppings. The seeds can then germinate and establish new populations. Without birds, the fruit would remain on the plant uneaten and eventually fall to the ground, and the next generation of trees and shrubs would not be planted at new locations, far from the parent plants.

In Thoreau's time, farmers had already recognized the costs and benefits of having birds in the landscape. They valued the insect-eating birds and despised the fruit-eating ones. Thoreau himself resented this purely economic way of valuing such fascinating and colorful creatures, writing:

When the protection of birds comes up, the legislature regard only the low use and never a high use; the best-disposed legislators employ one, perchance, only to examine their crops [a bird's throat pouch] and see how many [insect] grubs or cherries they contain, and never to study their dispositions, or the beauty of their plumage, or listen and report on the sweetness of their song. (*Journal*, October 15, 1859)

Human economic values of nature's gifts are ever-changing, depending on the moment, but the birds' place in nature is not determined by humans. It's at once both a liberating and a sobering thought, and one that Thoreau might have appreciated. On their own, birds are reacting to a changing climate, and in the process are now providing valuable insights to scientists and the broader human community.

Frosted elfin on lupine plant; photo by Ernest Williams.

*That interesting small blue butterfly . . . is apparently just out,
fluttering over the warm dry oak leaves within the wood
in the sun. . . . This first off-coat warmth just preceding the
advent of the swamp warblers brings them out.*

THOREAU, *JOURNAL*, APRIL 30, 1859

9. Bees and Butterflies

AFTER A WEEK OF UNUSUALLY WARM days and nights in early March, the crocuses are blooming in front of our house. As I enjoy the sight of their bright blue, yellow, and white blossoms against the background of matted-down grass, I notice the honeybees visiting the flowers. By the calendar it is still winter, but these bees have started their foraging several weeks early and found one of the only sources of food in the whole neighborhood. The bees and the flowers seem to be responding in the same way to the high temperatures.

Apart from some butterflies, most insects don't migrate. They have spent the winter under bark, deep in the soil, inside twigs, some as eggs or larvae, others as pupae. Some adult insects, like ladybird beetles, have spent the cold months hibernating in the leaf litter, and honeybees have spent the winter huddling in their hives. These insects need some combination of warm spring weather and lengthening days to stimulate their emergence from winter slumber. This makes them both susceptible to the harmful effects of climate change, such as late frosts, and potentially useful as indicator species.

Insects are a critical piece in the story of climate change. When birds return to Concord, and Massachusetts in general, they feed primarily not on plants but on insects. A few bird species, such as the ruby-throated hummingbird, drink nectar from the earliest blooms and sap oozing from broken bark, but most newly arriving birds feed on insects, such as flies, ants, and caterpillars, and other invertebrates, such as worms and spiders.

If my team and I hoped to use the events of springtime to show whether our climate is changing in Concord, it became obvious that we couldn't restrict ourselves to considering only the birds and plants.

Springtime is, after all, as much about the emergence of bees, butterflies, and a multitude of other insects as it is about flowers and birds. And insects are the connecting link, as bird feeds on insects, which in turn feed on plants. Given the critical role that insects play in ecosystems worldwide, it may be that our most important question will be this: What happens to *insects* if the climate warms up? We started our search for insect data in the fall of 2004, along with our general search for plant and bird data. In contrast to the abundance of plant and bird records, it took many years to find anyone who had bothered to record when insects become active in the spring.

The Helpful Honeybee

Honeybees were the obvious place to begin looking for useful data sets, and it seemed to us that the data should be abundant and readily available. Of all the insects that might be harmed by climate change, the type that matters most to humanity is the honeybee. Commercial agriculture depends on domestic honeybees to pollinate crops throughout North America and Europe, and wild *Apis* species are equally important throughout much of the rest of the world. In Massachusetts, honeybees are key pollinators of apples, blueberries, many fruit and vegetable crops, and numerous wildflowers. Also, because they begin foraging in the first warm days of spring, searching for the earliest flowers, honeybees should be useful indicators of a warming climate.

I expected, given how popular beekeeping has become, that there would be plentiful data from both commercial beekeepers, "sideliners" who kept bees as a sideline to make some extra money, and backyard hobbyists who keep a few hives for the fun and a few gallons of honey. Yet I discovered that data on the emergence of honeybees was all but nonexistent. Most beekeepers are aware of the connection between warmer temperatures and bee activity in the spring, but to date I've found no specific records that document exactly *what* days bees first are observed emerging from their hives.

In fact, when I called the president of the Massachusetts Beekeepers Association to ask whether his organization kept such data, he seemed to have trouble understanding what I was even asking. He explained that of course bees are more active on warm spring days; however, this was so obvious that he was sure that no beekeeper would ever actually write the dates down. He promised to ask his members if any of them had information on the dates for first spring flight, but he did not seem encouraging, and I never heard from him again.

In spite of these difficulties, I am sure that someday one or more citizen scientists will yet come forward with the observations needed for my research. I believe that somewhere out there, I'll soon find a science-minded beekeeper who can help me track whether (and how) honeybees' springtime behaviors have changed in recent decades.

The Forgotten Bees

While domesticated honeybees may have the highest profile because of their close connection with humans, there are bee species native to North America that might serve as harbingers of climate change. In contrast to honeybee colonies, which contain hundreds or even thousands of workers packed together in a single social unit, most native bee species are either solitary or communal, without the caste structure of highly social honeybees. There is one such solitary bee species that can be found in great abundance at the Sleepy Hollow Cemetery in Concord, where such famous Concord authors as Thoreau, Ralph Waldo Emerson, Louisa May Alcott, and Nathaniel Hawthorne are buried.

On the first warm days in April, a careful inspection of the sandy ground on south-facing slopes of the cemetery reveals one-inch-diameter mounds of newly excavated soil, with an entrance tunnel in the middle about the diameter of a pencil. At first glance it seems likely an ordinary ant mound, but the entrance is much larger, and a closer examination shows that each cavity is occupied by just one small bee. The bee, most likely an *Andrenid* species (known commonly as mining bees), can be observed kicking sand out of her nest during the excavation process. These bees live up to their name as solitary bees, with each bee, actually a female bee, acting individually and only building one nesting cavity for herself and her offspring. The female will later lay eggs in her tunnel and provision each one with food for the developing larvae. Such bees are not aggressive and are reported to sting only if caught and handled.

Thoreau urged botanists to become more familiar with the native solitary bees, because by observing these insects, botanists could learn when flowers are open and closed, and when they have nectar. Also, the bees are able to find even the most rare plants in the woods, a skill that many botanists pride themselves on, as Thoreau observed:

> *It is not in vain that the flowers bloom, and bloom late too, in favored spots. The tiny bee which we thought lived far away there in a flower-bell in that remote vale, he is a great voyager, and anon he rises up over the top of the wood and sets sail with his sweet cargo straight for his distant haven. How well they know the woods and fields and the haunt of every flower!* (*Journal,* September 30, 1852)

By the middle or end of April, all of the female bees at Sleepy Hollow Cemetery have excavated their nests, and the ground on the sunny slopes is covered with their mounds. In some areas, the nest mounds are edge to edge, and the entire surface is roughened with the tiny shovelfuls of soil kicked out by the bees. At this time of the year, the ground itself can appear to be rapidly shimmering; closer inspection reveals this shimmer to be masses of individual zigzagging bees, each flying a few inches about the ground, only rarely landing and never flying into the air. What is the explanation for this frenzy of activity? For solitary bees, this activity seems pretty social.

This is a mating swarm, during which females patrol their territory and fight off other females, sometimes wrestling with them. Male bees fly into these territories, find a willing female, and then land with her for a brief mating. A female that has already mated or has become exhausted retreats to her tunnel and rests. This frantic movement is the solitary bee version of speed dating, with lots of intense meeting, choosing, and mating on a compressed schedule.

It is clear that solitary bees have the potential to be a great climate-change indicator species, with bees probably emerging earlier in warm springs. However, we set this aside as a project for the future, as we were not able to find anyone who had monitored wild bee emergence in Concord or anywhere else in Massachusetts. However, gathering such data should be a research priority for North American scientists; the sunny, sandy banks so favored by mining bees and other solitary bee species, with their well-drained soils, are prime sites for new human homes.

We Shift to Butterflies

After six years of searching, we had still not located any data sets on spring emergence of insects. In the spring of 2009, Caroline Polgar, a Boston University graduate student, and I decided to make this a priority, and we began to invest time in e-mailing and talking with entomologists, trying to locate data that we were sure must be there but was eluding our grasp. And it turns out that butterflies would be the first group of insects to provide us with the data that we needed and would be major focus of several years of work.

In one of my initial meetings with Caroline, I told her she would be spending the next few years taking classes and teaching, but also wandering around in the Concord woods looking at plants and insects. To my delight, she told me that sounded like the ideal situation to her, and I think we were both reassured that she had chosen the right graduate program. Caroline recalled:

> Although I came into graduate school expecting to study plants, and I have, I have also explored other study systems, including butterflies. I had never thought much about the ecology of butterflies and knew very little about them, but once I was presented with the opportunity to investigate the effect of climate change on butterflies in New England, I jumped at the chance. It was quite intimidating to begin studying a group of organisms that I knew so little about, but I really enjoyed the opportunity. Through the Massachusetts Butterfly Club I got a chance to learn about the butterflies of my home state. A few years ago, armed with binoculars and field guides, Richard, his son Jasper, and I also struck out on our own to try and locate some of our study species in the Blue Hills Reservation. We were quite successful in finding and identifying several hairstreak species. When I started graduate school, I knew that I would meet many scientists and other students, but I never imagined that I would be working with local butterfly watchers, museum staff members, K–12 teachers, and student groups. My ability to explain my research to the public has improved tremendously, and I

don't feel that I am confined to the academic bubble that many graduate students feel stuck in.

For the purpose of our climate change work, butterflies seemed to offer a number of advantages over both honeybees and other insects. Unlike the vast majority of insects, butterflies are relatively large and brightly colored. Many have striking spotted and striped patterns in vivid blues, yellows, and oranges, allowing them to stand out against green foliage. Also, butterflies are easy to observe, as they glide and flutter in the open, stopping to feed on flowers. In the past, collectors often focused on butterflies when they were making insect collections. And now people are increasingly training their special close-focusing binoculars and digital cameras on these attractive, conspicuous, beautiful insects. We would search for butterfly data.

I felt that butterflies had considerable promise as climate change indicators, due to their association with sunny days. Also, Massachusetts has about one hundred native butterfly species, a number that made them a promising group for observation. A group of that size could provide some interesting contrasts and yet not overwhelm the enthusiastic naturalist with too many species to identify and remember. But I still did not know where to find any data on butterflies.

Sharon Stichter's Butterfly Data

We had been searching for insect data for six years and butterfly data for a few months when our luck suddenly changed with a single e-mail from a butterfly enthusiast and retired sociology professor. Sharon Stichter had heard of our work from the various postings and articles that we had written during our long search for data on the first appearance of birds and insects in Concord and Massachusetts. In the early spring of 2009, Stichter contacted us to see if we were interested in the data that she had been gathering for the past fourteen years on the first appearance of butterflies in her butterfly garden in Newbury, on the north shore of Massachusetts. When Caroline and I visited her garden a few weeks later, she also told us that she was part of a large community of people in Massachusetts interested in butterflies, who were all active in the Massachusetts Butterfly

Club. It was a joy to see her garden, as it was entirely composed of plants loved by butterflies, with lots of nectar-producing flowers and the kinds of leaves that caterpillars like to eat.

As we talked to her, we recognized that even if Stichter did not have exactly what we were seeking, the information that she had recorded so far in her garden represented the best data set on insects that we had yet identified. I also could see that her experience as a trained researcher and a member of the Massachusetts Butterfly Club meant that Stichter's data was likely to be highly reliable, with accurate identifications and a standard method of recording data.

What we were really hoping to find was someone who had been observing numerous butterfly species for at least thirty years, before the rapid temperature increases that began in the early 1980s. But we had yet to find that data, and perhaps it did not even exist. We remained determined to keep on looking for it, but in the meantime we would see what Stichter had to show us.

Since 1995 Stichter has been keeping a diary of what butterflies she sees in her garden. These are the records that Caroline and I began to analyze in the summer of 2009. Of the fifty-two species of butterflies and moths that Stichter has observed, sixteen species of butterflies and one species of day-flying moth are common enough that they can be analyzed statistically. The fifteen-year time period of Stichter's observations is too brief to reveal long-term trends in emergence dates over time, but Caroline and I could at least use her diaries to determine whether these butterflies are responsive to temperature. Once we analyzed all of the butterfly species from her garden together, we could see that these species of butterflies first appear at Stichter's garden about one day earlier on average for each 1 degree Fahrenheit increase in temperature. We were able to predict that over the past forty years butterflies have been appearing about four days earlier than in the past, as the eastern Massachusetts temperature has risen by about 4 degrees Fahrenheit. This response is about half of what we had seen for the flowering times plants in Concord and elsewhere in Massachusetts.

When we looked at individual species, we found that three notable species in Stichter's garden showed particularly strong responses to temperature, appearing earlier in warm springs. These are the cabbage white, the pearl crescent, and Peck's skipper.

The cabbage white was accidentally introduced into eastern North America from Europe about 150 years ago and is now one of the most common butterflies, frequently seen around gardens, visiting flowers and laying its eggs on cabbages, radishes, and other members of the mustard family. Its larvae do considerable damage to these crops. The pearl crescent is recognized by its orange color with black wing margins, and more specifically by the pale, pearly crescent on the underside of its hind wings. Peck's skipper is a small, stout-bodied brown butterfly that holds its relatively small wings at an angle; the males often wait on flowering plants for females to arrive. All three species have healthy populations in Massachusetts, and I wonder if their flexible response to temperature is one of the keys to their success. Perhaps this flexibility allows them to match the earlier leafing out of their food plants in warmer years.

After years of searching for insect data, Stichter's garden and her records represented the best resource we had located to that point on the impacts of climate change on Massachusetts insects at one location. The data are modest compared to the wealth of information that we have on plants and birds: just fifteen years of data on sixteen species of butterflies from a single garden. But as long as she continues her observations, Stichter's data will only grow more valuable over time, documenting the changing distribution of butterflies as changing climate allows more southern species gradually to become established in Massachusetts. As the climate becomes too warm for them, some of the more northerly species, for whom Massachusetts lies at the southern limit of their range, will gradually decline and disappear as they shift their distribution northward. We can already see these changes happening, and they can be expected to become more noticeable in coming years as the New England climate continues to warm.

The Great Northward Expansion

Butterflies are especially useful for tracking the changing distribution of species. Because many butterfly species are strong fliers, when their current habitat gets too warm, they can potentially disperse from their existing populations to colonize new, more favorable habitats. Butterflies are well collected and frequently observed by both amateur and professional lepidopterists, so we know the distributions of species with a good degree of accuracy. We might predict that butterfly species would be expanding more at the northern limits of their range in places like North America, Britain, and continental Europe, and that they would be found higher up in mountains than previously.

This is, in fact, exactly what scientists are finding. Many butterfly species are declining at the southern edge of their range, as the temperature is getting too hot for them or the habitat is changing, and they are expanding in at the northern edge of their range, as a few degrees of warming allows them to live in places that were formerly too cold.

Studies of butterflies in the Rocky Mountains have shown that many species live right at their temperature limits, and if you bring caterpillars a bit higher on a mountain where they are living, they will stop growing or freeze. However, as temperatures have been rising over the past few decades, the temperature envelope in which butterflies can live has also moved upward on mountains, allowing certain butterfly species to extend their ranges higher on the mountains.

At the other extreme are butterfly species that do not readily fly long distances; are found only in specific habitat types, such as sand dunes or native grasslands; and have larvae that only feed on a few rare plant species. These specialist species will be negatively affected by climate change because they are generally unable to colonize new sites elsewhere as fast as the climate is changing and will simply decline and go extinct. Climate change may occur so rapidly that it will cause winners and losers among butterflies, with certain butterflies becoming more widespread and others declining and disappearing forever.

It is worth pointing out that butterflies aren't able to home in on good territory when their current territory grows too warm. A butterfly living midway up a mountain does not deliberately fly upward on a mountain or farther north just because the temperatures have warmed by a few degrees. Suddenly confronted by a warming world, a population — be it insect, bird, mammal, frog, or reptile — will not deliberately move north to seek a more suitable patch of habitat. These animals cannot know that the period of warming is a long-term trend, or that temperatures will not soon return to normal. Also, they will not know that they need to head north to find cooler temperatures, and not some other direction, such as east or south.

Rather, the ability of a butterfly or any other species to discover new habitats often takes place simply because it's been a good year. In such years there may be an overabundance of healthy individuals, more than a location can support. Some individuals will strike out to find a new patch to call home. A high density of individuals can lead to fierce competition for scarce resources and mates, with the losers forced out. These butterflies will head out in all different directions — some north, some south, some east, and some west. Or if the species is on a mountain, some will head upslope and some will head downslope, and many will just follow the prevailing wind direction. The great majority of individuals will die without finding any new suitable place to live. A few will even travel farther into the present range of the species and may mate with other individuals of its own species. But a few butterflies, just by chance, will head farther north or higher in elevation. In the process, these particular butterflies may be lucky and settle down and establish a new population at a place that is now just warm enough for the species but one or a few years ago was too cold. This process of colonization is more likely to occur when the population is doing well and there are many healthy individuals dispersing from the colony. A population already under stress from high temperatures may already be in decline and not produce enough healthy individuals that can leave the locality and establish new populations.

The Massachusetts Butterfly Club

During our first conversations with Sharon Stichter during the summer of 2009, she kept referring to the activities of the Massachusetts Butterfly Club, something that we had never heard of. As we learned to our surprise, butterflies are the only major group of insects in Massachusetts with its own dedicated club, in this case composed of butterfly enthusiasts. Many members of the Massachusetts Butterfly Club are former birders who joined looking for new challenges. In addition to regular meetings, the club organizes outings and publishes checklists of species observed in different localities throughout the state.

I realized that the club might be the key to our search for insect data. After the frustration of not being able to find any data for honeybees or solitary bees, despite months of looking, Caroline and I allowed ourselves to feel cautiously optimistic about the possibility of unearthing a butterfly data set. Could the members of the Massachusetts Butterfly Club provide us with extensive observations of flight times? If they could, butterflies might prove to be our sought-for indicator group.

Here is where we got very, very lucky. As Stichter explained to us during several conversations, the Massachusetts Audubon Society decided in 1986 that they wanted to expand their traditional focus on birding to include butterflies. For this project, they decided to organize a group of serious birders, train them to identify butterflies, and carry out a systematic inventory of Massachusetts to determine what butterflies were present in Massachusetts, where they were found, what season they were active, and what the adults and caterpillars were feeding on. The survey was carried out for five years, and the results form the basis of the *Massachusetts Butterfly Atlas*. The survey documents include thousands of observations made each year along with the dates and places of observations for all Massachusetts butterflies.

And the story gets better. After the *Butterfly Atlas* project finished, the contributors decided to form the Massachusetts Butterfly Club to continue making their observations. With over one hundred members, this club has been making observations on butterfly occurrences from throughout Massachusetts . These modern butterfly enthusiasts are much less likely to collect specimens than past lepidopterists, as they are concerned about causing further declines in species in which population numbers are sometimes already declining. And many club members don't want to catch and kill the butterflies that they love so much. The club members are more likely to simply record their observations in journals and use digital cameras to "capture" a butterfly; these photographs records are later posted on the Massachusetts Butterfly Club website. We also discovered that Stichter, the woman whose butterfly garden we had visited and we were already working with, was the club archivist and had all of the club's data on electronic spreadsheets. These spreadsheets contain tens of thousands of observations from across the state.

Caroline and I could see right away that these modern records of flight times from 1986 to the present time could be analyzed to determine species responsiveness to variation in current climate. We also wondered if we could locate enough museum specimens of butterflies collected from

Massachusetts in the hundred years prior to 1986 that we could use to compare with the club records. We were eager to see if butterflies are now being observed by Massachusetts Butterfly Club members earlier than when they were collected by past lepidopterists. The place to begin was the Museum of Comparative Zoology at Harvard University, where they have the largest natural history collections in New England. Stichter explained:

> I am always awestruck when I enter the Harvard butterfly museum. It is like a temple devoted to butterflies, yet collectors had to kill hundreds of thousands of butterflies to fill the museum drawers. For me, the thrill of butterflies is watching them glide through the air, sip nectar at flowers, and deposit their eggs on suitable plants. I go to the field with binoculars and a camera, but in the past butterfly people carried only a net to capture these aerial insects. Even though we no longer collect butterflies, these old collections still provide us with so much necessary information to understand butterfly ecology and conservation.

Elfins and Hairstreaks: A Natural Experiment

In early 2010 we decided to start work in a serious way on butterflies. However, which species should we study? Not all species were suitable. We needed to find species that were reasonably common and had a relatively brief flight period. After talks with Stichter and the butterfly research group at the University of Massachusetts, we decided that the ideal group of butterflies might be elfins. Small brown butterflies in the genus *Callophrys* with wingspans of about one inch, these butterfly species are typically found at the edge of forest habitats. They are members of the gossamer-winged butterflies or Lycaenidae family. This family is notable because the caterpillars of this family are often "farmed" by ants for the sweet fluids they secrete from special glands on their bodies. The ants in return protect the caterpillars from attack by other insects.

Elfin butterflies offer distinct advantages for climate change research. They appear in the early spring and so are likely to have their emergence times primarily determined by temperature rather than day length. Also, they are active for a short duration, with one generation lasting just a few weeks, unlike many other butterflies that can have staggered emergence times, lifetimes lasting several months, or even two or more generations per season.

In Massachusetts there are five elfin species that are common enough to study. The eastern pine elfin is one such species. It has a band of small dark crescents, each edged with silver, on the reddish-brown undersides of its wings. Pine elfins emerge in the early spring and feed on the nectar of clover and other flowers. The males wait around in pine trees for the females to show up. After mating, the females lay eggs on young pine needles, which supply the caterpillars with their main source of food. Another elfin species that appears in the spring is the frosted elfin, recognized by a band of silvery hairs on the outside of the hind wing and a short extension to the hind wing called a "tail." Caterpillars of the frosted elfin eat the flowers and seed pods of lupines, false indigo, and other legumes.

Elfins appear in the spring because they overwinter as pupae. When conditions are suitable in the spring, either because of warmer temperatures or longer days, the butterfly emerges from its cocoon and begins to fly. Females lay eggs in the late spring, and the caterpillars feed on leaves during the summer and pupate in the fall.

A related set of five lycaenid species in Massachusetts, the hairstreaks of the genus *Satyrium*, has a different pattern of development than the elfins, providing an interesting contrast for climate change research. Hairstreak butterflies appear only in summer, whereas elfins appear in spring. The reason behind these different emergence times is that hairstreaks overwinter as eggs; the eggs hatch in the spring, and the small caterpillars feed on young leaves during the first months of the growing season. The caterpillars form their cocoons during the late spring and early summer, and the hairstreak butterflies only emerge in the height of summer, often in late June, July, and early August. The adults feed on floral nectar, mate, lay their eggs, and die in the last days of summer. And so the cycle continues.

In the fall of 2010, Caroline Polgar and I began to form a working prediction that the emergence of elfin species in the spring is to some extent controlled by warming temperatures because they are active in the spring. In contrast, butterflies active in late summer, like the hairstreaks, would be less affected by temperature because they emerge from their cocoons when temperatures are already warm enough for their activity; it is likely that the emergence of hairstreaks would be more affected by day length or rainfall rather than temperature. We would further predict that elfins in particular would be emerging in the spring earlier now than they were decades ago. The two groups of species represent a natural experiment for testing the impact of climate change on insects. We had data from the past thirty years from the Massachusetts Butterfly Club to begin our investigation, given to us by Stichter, but we were especially interested in seeing if there was even older data on butterfly flight times. Could we find data going back one hundred years or more, perhaps even to the time of Thoreau?

It turns out that data set for this natural experiment is scattered across New England and the rest of the country, as insect specimens in natural history museums and the research collections of colleges and universities. Housed in cabinets with long, flat drawers are tens of thousands of carefully preserved moths and butterflies, used by lepidopterists and biologists studying everything from evolution, classification, animal distribution, and plant-animal interactions. Almost every specimen, no matter how old or where in the world it was collected, bears an identifying label giving the date it was collected, where it was collected, and who collected it. For butterflies, a date of collection provides at least one record of the flight time for that year; the flight time for a species is the period of time from the first emergence of a species in a year until the last individual dies at the end of the season. We eventually found fourteen institutions housing natural history collections containing Lepidoptera collected in Massachusetts. For the past 120 years, naturalists in Massachusetts and visitors to the state have been intensively collecting thousands of butterfly specimens. The butterflies are netted, then put in a killing jar. Later the wings are carefully spread open and the insect is pinned flat to prepare it for the display case. Periodically experts will examine pinned museum specimens to identify them to species, if possible. Certain butterflies in museum collections are only identified to the level of their taxonomic family and can remain unidentified for years until they are examined by scholars who can make an authoritative identification.

Caroline and I decided to begin with the collections closest to hand, those at our own institution, Boston University, and the enormous collection of butterflies and moths housed across the river at Harvard's Museum of Comparative Zoology, where the de facto curator of butterflies was for some years the Russian-born novelist Vladimir Nabokov, author of *Lolita*, and where modern work on Lycaenid butterflies is carried on by Professor Naomi Pierce. We gradually extended our search to include Massachusetts specimens held by other museums across New England and the rest of the country, and even across the Atlantic at the British Museum. By contacting and sometimes visiting the natural history museums that have specimens collected in Massachusetts and recording the label information, we built up an extensive picture of days of the year that elfins and hairstreaks have been flying in past years. Many of the museums are in the process of photographing all of their specimens along with the labels and putting this information online, making the digital collection available to any scholar with an Internet connection.

To accomplish the work of locating specimens and formulating research questions, we added some additional butterfly experts to our team: Ernest Williams from Hamilton College and Colleen Hitchcock from Boston College. Sharon Stichter proved to be an especially dedicated member of the group, spending long hours extracting label data from butterfly specimens, as she related:

> In order to understand our future, we must first understand our past, and one aspect of this climate change study that appealed to me was our effort to review all the museum specimens of the focal species, which in the case of New England stretch back over one hundred years to the mid-1860s. We had to pull together specimen-label data from many different museums and collections to get a complete picture. I enjoyed this so much that I still volunteer at the Harvard Museum of Comparative Zoology.

We were hoping to find at least 50 Massachusetts museum specimens of each of the 10 common elfins and hairstreaks. In the end, we found around 750 specimens for these 10 species, around 75 specimens per species. This was not a huge number, but we thought that it might be just enough. But there was an interesting twist to the collection dates. Almost all of these specimens had been collected from 1880 to 1980, with very few specimens collected during the past thirty years, from 1980 onward. This is where the Massachusetts Butterfly Club data is so crucial. From

1986 to the present time, the Massachusetts Butterfly Club members have made over 4,300 observations of the flight times of these same species. By comparing these modern records with the data from museum specimens, we could potentially determine if butterfly flight times have been changing over time in Massachusetts.

When we look at the whole period for which we have data, from 1893 to the present, we find that hairstreaks are now flying earlier now than they were in the past, whereas elfins are not, which is contrary to our prediction. However, for the long period from 1893 to 1985, we have very few or no observations in many years, which tends to weaken the analysis. When we just look at the most recent three decades using the Massachusetts Butterfly Club data, we can see that both hairstreaks and elfins are having earlier flying seasons over time, with elfins flying six days earlier over this time period and hairstreaks flying three days earlier.

Warming temperatures are clearly contributing to this earlier flying. For nine of the ten species, temperatures in the two months before the flying period influence flight times, with one to two days earlier flying times for each 1 degree Fahrenheit increase in temperature. The first three months of 2012 were the warmest ever recorded in the Boston area, with temperatures about 9 degrees above normal. It is not surprising that this year set many new spring records for first observations for brown elfins and frosted elfins, and many other butterflies in different families.

Our analysis also included collection location, as we expected that butterflies in warmer locations of Massachusetts would emerge earlier than those in cooler locations. The climate of Massachusetts ranges widely, from the coastal islands of Martha's Vineyard and Nantucket, with the moderating influence of the sea, to the warm lowlands of eastern Massachusetts with Boston at its center; and from the cooler Berkshire Mountains to the west side of the Connecticut River Valley to the central hill country north of Worcester, centered on Mount Wachusett.

How do we encompass this great climatic variation in a relatively small geographic area? The U.S. Department of Agriculture has provided a simple solution through their plant hardiness zones. Massachusetts has been divided into six zones for the purposes of agriculture and horticulture on the basis of annual temperature cycles. And for our purpose, each of the butterfly observations can be assigned to one of these hardiness zones, to determine both how location in Massachusetts and changing climate together predict the window of flight times of these butterflies.

What we found was that location was an important factor along with annual variation in temperature in determining flight times. Each year elfins tend to fly earlier in coastal locations of eastern Massachusetts, as these regions are the first to emerge from the embrace of winter. In contrast, hairstreaks fly earlier in inland regions, away from the moderating influence of the ocean, where summers tend to be warmer.

In an interesting twist to this story, while we were working on our data analysis, the U.S. Department of Agriculture issued new maps of plant hardiness zones for the United States, with many of the zones of New England being reclassified to the next warmer category. The old USDA maps were widely considered to be out-of-date because they did not account for the warming temperatures that were occurring across the country. Farmers, gardeners, and landscape designers all knew that the old plant hardiness zones did not conform to the new reality of climate change. The vigorous growth of fig trees, southern magnolias, and other warm-climate species in Boston had already demonstrated the changing climate to a careful observer. And finally the USDA had to acknowledge the reality of warming temperatures with new maps, even though political considerations prevented them from saying it was due to climate change.

For our studies, we just used common species of elfins and hairstreaks and changing flight times to look for evidence of climate change. However, about a year after we started our butterfly work, another couple of butterfly researchers at Harvard University, Elizabeth Crone and Greg Breed, also started working with Sharon Stichter and the Massachusetts Butterfly Club data to determine if the entire butterfly community of Massachusetts was changing in abundance due to a warming climate. Their prediction was that warm-weather-loving southern species would be increasing in abundance in Massachusetts over the past twenty-seven years and cold-loving northern species would be declining in abundance. And this is what they found. Of twenty-one distinctly northern species, such as the Atlantis and Aphrodite fritillaries, seventeen were showing significant patterns of decline over time. Almost all of the increasing species were southern species, such as the giant swallowtail, the Zabulon skipper, and the frosted elfin.

At the end of our first year of analysis, we presented our results to a meeting of the members of the Massachusetts Butterfly Club. They were clearly excited that the club's data was being used to address a contemporary scientific topic. This was an additional benefit of this project for us. Not only were we able to discover a great new data source on insects, but we were able to energize the members of the butterfly club and to demonstrate the great contribution that citizens can make in climate change research.

In the end, we had found the insect data set that we had been looking for and that we could compare with our bird and plant studies. What we learned from all our studies viewed together was that plants in Massachusetts are responding to climate change by flowering about 1.7 days earlier for each degree Fahrenheit increase in temperature, which is indistinguishable from the average of 1.7 days earlier flight times for each degree Fahrenheit for elfins and 1.6 days for hairstreaks. Right around the time we were getting ready to submit our article, a study was published on the effects of climate change on wild bees of the eastern United States; quite strikingly this study showed that bees fly 2.0 days earlier in the spring for each degree Fahrenheit warmer temperature; this is pretty much the same as what we found for butterflies and plants. In contrast, our previous studies had shown that Massachusetts birds only arrive 0.4 days earlier on average for each degree Fahrenheit warmer temperature. The overall conclusion is that plants, bees, and butterflies are responding strongly and in a similar way to a variable and changing climate, but that migratory birds are much less responsive.

These results with a few groups of insects were very exciting, but we still needed to know if they would hold up for other groups of insects as well. We needed to know if our results with butterflies were part of a more general pattern, and our eventual goal was and still is to determine how these changing patterns of insects influence birds and plants.

*Fishing in Concord; Harold Wendell Gleason photo 1905.3 taken on
June 10, 1905, at Red Bridge, Concord, Mass. Concord Free Public Library.*

The air over the river by the Leaning Hemlocks is filled
with myriads of newly fledged insects drifting and falling as it were
like snowflakes from the maples, only not so white.
Now they drift up the stream, now down, while the river below is
dimpled with fishes rising to swallow the innumerable
insects which have fallen into it and are struggling with it.

THOREAU, *JOURNAL*, JUNE 15, 1850

10. From Insects to Fish to People

THOUGH WE CALL THEM MAYFLIES, the aquatic insects in the order Ephemeroptera in fact emerge anytime between March and September, depending on the species. Beginning in March, in ponds and lakes across North America, they begin to emerge from the water in their adult winged forms. The mayflies often emerge on a windless day, and it can look as though rain is falling on the pond as the surface is rippled by each new individual breaking the surface and taking flight.

They have spent the winter as nymphs in the rocks and mud of the bottom, feeding and growing, before beginning the last stage of their metamorphosis. Thousands, sometimes even millions of adults of one species can emerge in a single night. Mayflies are ephemeral, as their Latin names suggests, adults living only long enough to find a mate, and many not that long. As the mayflies emerge, the surface of the water is roiled by fish, in a frenzy to consume mayflies as rapidly as possible, and the air above the pond is crisscrossed first by swallows and swifts, and as dusk falls by bats, which also take advantage of the emergence for a huge feast.

First out, in March, are the blue-winged olives. In April the blue quills and mahogany duns emerge. In May the sulphur duns, the gray foxes, and brown drakes emerge in their own spectacular shows. Over the rest of the summer, flight after flight of gossamer-winged insects takes to the air, until the last of the mayflies emerge in September.

Other insects are emerging too: caddis flies, stone flies, dragonflies, midges, gnats, crane flies, and a whole host of other insects participate in this parade. Why should mayflies and other aquatic insects emerge in these huge numbers on just a few days or evenings in the year? Science offers a two-part explanation. First, by emerging suddenly en masse, even though many individuals will be eaten by the waiting predators, the insects' numbers are so huge that most will escape. Second, by emerging together, they will readily find mates and be able to lay fertilized eggs to form the next generation. This is especially important for species like the mayfly that don't have functioning mouthparts as adults and live only one night.

Fish such as trout and bass are attuned to this progression and focus their attention on catching and eating insects during the vulnerable time that the insects leave the relative safety of the rocks and mud of their larval stage and rise to the surface to emerge as adults. The fish form specific search patterns to catch each type of these insects as they become abundant for a few weeks, or perhaps even just a few days or nights, and then finish for the year.

Thoreau recorded the activity of such an emergence, perhaps of mayflies, in his journal of June 9, 1854:

> The air is now pretty full of shad-flies, and there is an incessant sound made by the fish leaping for such as are struggling on the surface; it sounds like the lapsing [falling] of a swift stream, such among the rocks. The fishes make a business of thus getting their evening meal, dimpling the river like large drops as far as I can see. Meanwhile the kingfishers are on the lookout for fishes as they rise.

For hundreds and probably thousands of years, fishermen have known that their chance of catching fish can be increased if they use lures that imitate in appearance whatever type of prey the fish are currently eating. By the 1700s, advances in angling equipment were enabling fishermen to pursue freshwater fish by fashioning lures, especially "flies" that resembled the adult forms of aquatic insects. Long, flexible fishing rods with lightweight lines allowed these anglers to cast a line into the very place where fish were feeding. By moving the lure around in just the right way, the fisherman could give the artificial fly the appearance of an emerging insect trying to wriggle free. In this way, fishermen have learned to catch a fish based on what insects are available at that exact time and place where the fish are feeding.

One evening in 2007, after a public lecture I had just presented in Boston on my climate change work, a man told me that fishing lodges kept records of when different insects emerge in order to be able to use the correct lure. Of course! What a great new potential source of data for climate change research. That night I could barely sleep I was so excited about the new possibilities awaiting me. When I woke up in the morning, I was certain that somewhere out there I would soon find a fishing lodge, a fishing club, or government agency that would have recorded when such insect emergences take place every year. I would then be able to use these records as indicators of climate change. My first calls did not turn up anything, but I was not discouraged. I spent the next few days calling dozens of people around New England and then the rest of the country, and sending out e-mail messages to every place I could think of. I was optimistic, thinking that I was on the verge of a big breakthrough. In my dreams, both waking and sleeping, I found myself talking to someone who told me that they had fifty years of records of every emergent insect that has been seen at a famous fishing site.

Warming Water at Walden

I began by contacting fishery biologists with the U.S. Fish and Wildlife Service and state fisheries agencies. They were quite interested in my questions but didn't have any information to share, nor did they have the slightest idea where such information might exist. There were certainly plenty of "hatch charts," showing when various insects emerge in different rivers and lakes, but these were based on typical years and were not linked to annual records, as far as I could determine. These biologists also knew that warming water temperatures alter stream ecology, often in ways that are detrimental to fish that thrive in cold water, such as trout and salmon. But they haven't yet undertaken research to see how such warming will affect the timing of insects on which the fish feed.

Thoreau also understood the relationship between temperature and the distribution of fish. On August 22, 1860, he decided to measure the temperature of the water at the bottom of Walden Pond, one hundred feet below the surface. In his journal, he describes lowering a bottle of water on a line and leaving the bottle there for thirty minutes. Hauling the bottle up, he used a thermometer to measure the water temperature, getting a reading of 53 degrees, which was 22 degrees lower than the surface temperature. He was surprised by this range of temperatures:

What various temperatures, then, the fishes of this pond can enjoy! They re-
quire no other refrigeration than their deeps afford. They can in a few minutes
sink to winter or rise to summer. How much this varied temperature must have
to do with the distribution of the fishes in it. The few trout must oftenest go
down below in summer.

Walden Pond has likely undergone a general warming since Thoreau's time, leaving a smaller volume of water for the trout. On July 7, 1860, Thoreau measured the temperature of Walden's water at a depth of four feet and recorded a temperature of 71 degrees. For the past five years, I have repeated these measurements each July and have found that the temperature has varied from 76 to 84 degrees in different years. Even though Thoreau only measured the pond at one place in one year, this is an indication, along with the earlier ice-out described in chapter 2, that Walden has undergone a general warming over the past century and a half.

After talking to all of these government officials about aquatic insects without success, next I tried the people at Trout Unlimited, a large conservation organization dedicated to protecting fish habitats, as well as individual fishing lodges throughout New England. In the end, after several days and hundreds of telephone calls and e-mails, I was unable to find any yearly records of emergence dates or any idea where to find such data from New England or anywhere in the northeastern part of the country. I had nothing to show for my time other than general hatch charts for typical years, yet I still bet that somewhere out there is a methodical fisherman who fishes every day or a couple of times a week in the same place and who has kept a detailed journal noting when the mayflies and other aquatic insects emerge from the water. Or perhaps this person might be a thoughtful guide who realizes that he might increase his worth to his clients by having more accurate predictions of when the fish might be eating certain insects. But I have not yet found this person or persons, whoever they may be and wherever they live. I am sure that someday I will find them. Or perhaps some young fisherman will read this passage and start the type of journal that will provide this information to a future scientist.

I would especially like to suggest Walden Pond as a site for collecting new observations. Each spring Walden Pond has one such dramatic emergence in which aquatic larvae are transformed into winged adult insects.

Early each April, over the course of a few days, the shore and paths around Walden are filled with swarms of small black midges, probably in the genus *Chironomus*. Recording the emergence time of midges would make a valuable contribution to our understanding of the impact of climate change on Walden Pond.

The midges hover in the open air about four to six feet above the paths of Walden Pond. This height presumably allows individuals in the swarms to meet and mingle without the inconvenience of flying around shrubs and tree seedlings. These tiny insects spend their brief lives out of water searching for mates. If they are successful, the females will lay their eggs on the surface of the pond. When the eggs hatch, the larvae emerge to build tubes of mud on the bottom of the pond. Known as blood worms because of their reddish color and shape, the larvae live in their mud-tube nests, feeding on microorganisms, and are in turn fed on by the larvae of other insect species as well as by small fish.

Midges in the family Chironomidae aren't biting flies, like the fearsome black flies or deerflies. However, city-dwellers readily mistake these Walden midges for mosquitoes and often become alarmed by their large numbers.

The Value of Fishing

When I walk around Walden Pond, I sometimes see dozens of fishermen, all trying to catch some of the nine hundred or so trout released into the pond each year by the Massachusetts Department of Fish and Game. What brings them to the shore, fishing gear in hand, is the fishing, not the fish; most hobby fishermen don't catch a single fish when they go out for a few hours. They just enjoy being outside, away from their workaday cares. Many also enjoy talking with their fishing buddies. Some fishermen bring their children with them, getting them out of the house, away from electronic distractions and closer to nature. In the process, both the fishermen and their children may slow down the pace of life and enjoy the simple act of being alive. Slowing down allows them to see birds, insects, and wildflowers and so learn more about the natural world.

In Thoreau's time, fishing often was still undertaken for subsistence. People went fishing to catch their dinner, something to eat along with cabbage and potatoes. Thoreau believed that fishing and hunting were important in getting people out of their homes and into natural places. Even in Thoreau's time, people were moving off the farms into the new towns of the Industrial Revolution and becoming separated from the sights and sounds of nature. Over time Thoreau hoped that some of these fishermen and hunters would transition to the purer forms of nature appreciation. He notes:

> *[A young man] goes thither at first as a hunter and fisher, until at last, if he has the seeds of a better life in him, he distinguishes his proper objects, as a poet or naturalist it may be, and leaves the gun and the fish-pole behind.* (*Walden*, 232)

Thoreau believed that time spent outdoors hunting, fishing, and cutting wood allowed young people to grow in their knowledge of the natural world, beyond what they could learn from formal study:

> *His [the outdoorsman's] life itself passes deeper in Nature than the studies of the naturalist penetrate; himself a subject for the naturalist. The latter [the naturalist] raises the moss and bark gently with his knife in search of insects; the former [the outdoorsman] lays open logs to their core with his axe, and moss and bark fly far and wide.* (*Walden*, 308)

Modern fishermen still visit Concord for a few hours of fishing, but now they are joined by hundreds of people enjoying the beach to sunbathe and swim, two recreational activities largely unknown in Thoreau's time. Thoreau would have been even more surprised by the tens of thousands of people who visit Walden Pond each year to walk from the parking lot to the site of his cabin. Many of them will drop a pebble on the stone pile next to the foundation, attempting to forge a link to Thoreau's time.

Margaret's Approach to Fishing

By now you may be asking yourself, *What does all have to do with climate change?* Well, if spring events are happening earlier and aquatic insects are hatching and emerging as adults earlier in the spring, this has implications for fish every bit as much as it has for birds, as many fish eat both insects *and* their larvae. Changes in temperature could adversely affect the availability of larvae and adult insects on which fish depend for food. Changes in water temperature could also directly affect the ability of fish to survive and grow; many fish, such as trout, are very sensitive to water temperature and even slightly warmer temperatures could force certain fish to deeper water or even kill them. Also, warmer water has less dissolved oxygen than cold water, and the lack of oxygen can cause certain fish to suffocate in warm water. And while recreational fishermen in Massachusetts may be able to go home empty-handed and pull some chicken from the fridge for dinner, this isn't the case for millions of impoverished people around the world.

As I watch fishermen around Walden out to catch recreation as much as fish, I cannot help but contrast them with my wife. Margaret has a very different approach to fishing, one more in tune with the rural poor both in the United States and the developing world. Margaret grew up on a subsistence farm carved out of the rain forest of Malaysian Borneo, when it was still a British colony. Her family lived near the great Baram River, which was filled with both fish and crocodiles. Margaret's family was so poor that there frequently wasn't enough food to eat, and they were forced to search the nearby forests and riverbanks for edible plants and wild game. And they fished. They didn't fish for recreation; the fish they caught were an absolutely crucial supplement to their minimal diet of rice and vegetables. Their fishing gear consisted of a short length of fishing line tied to the end of a pole, with a bent sewing needle for a hook. A live grasshopper or cricket was the usual bait, though a small berry might serve as well. This was serious work, as failing to catch a fish might mean almost nothing to eat for dinner for her parents and their eleven children. Margaret's experiences in rural Borneo half a century ago are still true today; in coastal areas of the developing world, more than 50 percent of people's protein comes from fish and other seafood. And to further emphasize the importance of fishing for poor people, 97 percent of the people who catch fish for a living reside in the developing world, according to the Food and Agriculture Organization (FAO) of the United Nations.

Margaret brings her same practical approach to her fishing in New England. When we were married three decades ago, our first fishing excursions were to Lake Sunapee in New Hampshire. The water was clear and cold, fed from the melting snow of nearby Mount Sunapee. The lake had beautiful landlocked salmon, but there were so few of them, and so many fishermen, that only the most dedicated fisherman could count on catching one or two over the whole summer. Margaret found this kind of fishing unfathomable: spending days on the water with nothing to show for it and no fish to bring home for dinner! Fishing in Walden Pond she likewise regarded as wasted time, because there are so few fish to catch.

But then we discovered Perkins Pond, only a few miles from Lake Sunapee, a hidden gem in which the water was enriched with nutrients flowing in from nearby maple swamps. As a result, the lake teemed with life. It was filled with frogs and tadpoles, turtles, mayflies, dragonflies, and other aquatic insects, and, most importantly, plenty of fish, especially sunfish and bass. Most of the fishermen who visited the lake pursued bass and caught them with some degree of success. Margaret's target was the common pumpkinseed sunfish, a yellow-brown fish, oval in outline, that reaches about eight inches long and a bit more than half a pound in weight. This is an active fish, not like the sleepy bass that rests during hot days. Swarms of sunfish remain active throughout the day and compete for the first bite of a baited hook. So in just an hour or two, Margaret could easily catch a dozen or more fish, enough to feed our whole family.

When we went saltwater fishing, Margaret again wanted to make sure we had something to eat at the end of the day. On a visit to Martha's Vineyard off the coast of Massachusetts, all the other fishermen along the rocky breakwater were going for striped bass with fancy saltwater rods and lures. Margaret used her regular freshwater fishing rod; baiting the hook with sandworms, she caught small fish among the rocks. And at the end of the day, we had fish for dinner, while the other fishermen casting for stripers had nothing to bring home. If they wanted to eat fish, they would need to head to a seafood restaurant for dinner.

On another day of fishing at Martha's Vineyard, the water was so filled with tiny shrimp that the water seemed to be filled with pinkish streamers, waving in the currents. This was the abundant type of small shrimp known generally as krill that forms a key building block in many cold-water marine ecosystems. Ever inventive, Margaret caught several pounds of krill in just a few minutes with a child's dip net. That night she fried the krill in olive oil and then added in some soy sauce for flavor. This formed a high-protein companion to our rice, though I have to admit the krill did not taste like much beyond the soy sauce and olive oil. We later read that the sharp shells of the krill can irritate some people's intestines, leading to stomach upset, but we suffered no such ill effects.

Eating Smaller Fish and Krill

Margaret's fishing activities show on a small scale a way to address the world's overfishing problems. Instead of eating large predatory fish, such as tuna and swordfish, which eat small fish that eat krill, we would be far more efficient if we just ate the krill directly. Krill are already being harvested from the ocean on an industrial scale, but they are mainly being used as fish food on fish farms for more desirable fish, such as salmon, tilapia, and catfish. The same practice of harvesting wild krill as feed for farmed fish applies to a number of small fish species, such as anchovies and herring. This inefficient process uses several pounds of high-quality krill or fish protein to make one pound of large farmed fish protein. If we ate the krill and these various small wild fish directly and reduced our consumption of farm-raised fish, we could reduce our overall catch of krill and many wild fish species and still have more seafood to eat. Of course, shifting to harvesting krill in international waters would require establishing a well-managed fishery with all fishing nations agreeing to act cooperatively. The record of the community of nations acting together to protect valuable marine resources has been mixed so far, with sharks, swordfish, tuna, many sea turtles, and certain whales continuing to decline.

Thoreau appreciated the value of such small fish to the diet of poor rural people, noting:

> The river, too, steadily yields its crop. In louring [dark and threatening] days it is remarkable how many villagers resort to it. It is worth more than many gardens. I met one, late in the afternoon, going to the river with his basket on his arm and his pole in hand, not ambitious to catch pickerel this time, but he thinks he may perhaps get a mess of small fish. These kind of values are real and important, though but little appreciated, and he is not a wise legislator who underrates them; . . . And who can tell how many a mess of river fish is daily cooked in the town? They are an important article of food to many a poor family. (Journal, July 20, 1851)

It may seem like a stretch to connect aquatic insects and hunger when you live in Greater Boston, where there's a supermarket nearby. But my wife's family history makes me especially conscious of how climate change has the potential not only to disrupt ecological relationships among insects and birds, but also to affect food resources needed by poor people living in rural areas throughout the world. According to the Food and Agriculture Organization, an estimated 852 million people do not have enough food to eat. The loss of insects — along with birds, plants, and fish — will have real negative effects on these people. If we start losing insects due to climate change, the way we've started losing plant, bird, and fish species — and it seems likely that this has already begun — we'll be losing much more than annoying buzzing creatures that invade our summer evenings. Without these pollinators, the produce section of the supermarket would be missing many of the fruits and vegetables whose variety and plenty we take for granted: from apples, cherries, and melons to tomatoes, cucumbers, and beans. We'll be losing detritivores that break leaf litter down into nutrients, allowing forests to grow. And, yes, we'll be losing a crucial link in the food web that feeds fish and ourselves.

Climate change will affect insects in another way, one that could have serious implications for human health beyond meeting our daily need for calories. Changing climate could trigger increases in the mosquito populations and other insects that carry human diseases.

Mosquito trap and child covered with netting in a swamp in southeastern
Massachusetts; photo by Richard B. Primack.

Would it not be a luxury to stand up to one's chin in
some retired swamp for a whole summer's day, scenting the sweet-fern
and bilberry blows, and lulled by the minstrelsy of the gnats
and mosquitoes? . . . Say twelve hours of genial and familiar discourse with
the leopard frog. The sun to rise behind alder and dogwood, . . .
and finally to sink to rest behind some bold western hummock. . . . Surely,
one may as profitably be soaked in the juices of a marsh
for one day, as pick his way dry-shod over sand.

THOREAU, *JOURNAL*, JUNE 16, 1840

11. Clouds of Mosquitoes

IT'S A SOUND THAT, along with the slam of a screen door and the sound of a lawn mower, unmistakably signals the return of warm weather: the whine of the mosquito. We generally take notice of mosquitoes because they hover around us, land on us, and bite us. Their bite feels like getting poked with a needle and leads to an itchy bump that lasts for hours or days. We may be vaguely aware of their habitats, but only as part of the effort to avoid them. But they're more than mere annoyances — as we're increasingly aware, mosquitoes are vectors for a range of diseases, ranging from malaria and yellow fever to dengue fever to West Nile virus to Eastern equine encephalitis. Mosquitoes are the most dangerous animals on the planet in terms of their ability to not only kill but seriously debilitate humans. Malaria alone affects over a quarter billion people worldwide and kills approximately 1 million people each year; it is estimated that one in five children in Africa will die of malaria. In South America and Africa, there are around 200,000 cases of yellow fever each year, resulting in 30,000 deaths. Despite the availability of an effective vaccine for yellow fever, the number of cases of this mosquito-borne disease is now increasing due to political instability in many countries. And mosquitoes carrying these diseases are among the insects most likely to actually benefit from climate change, which has worrisome implications for human health.

If you're familiar with the mosquito as a tropical pest, you might think that mosquitoes are among the least suitable subjects for climate change research. In some areas of the tropics, that would be true — they buzz and bite all year long in warm climates, with few behavioral changes to help track alterations in seasonal emergence. But in the northernmost parts of the United States, including Massachusetts, mosquitoes are seasonal pests, present from the frost-free days of early May to sometime in September or October, when the first frosts kill them. These insects overwinter as eggs or larvae (depending on the species) in marshes, ponds, bogs, and other bodies of water. If the water becomes too cold in the fall, the larvae can't finish their development and remain dormant, and the adult mosquitoes can't emerge until the ice on the surface melts in the spring. At the end of the growing season, any remaining adult females are killed by the cold. For such species, a warming climate could mean a longer flying — and breeding — season, and a larger population size.

Hard as it might be for us to appreciate, as we slather on repellent, install screen doors, and invest in bug-zapping devices, Thoreau did not share our modern distaste for mosquitoes. Rather, he celebrated their high-pitched whine as a sound of wild nature. In Thoreau's time, the connection between mosquitoes and disease was unknown, and Thoreau admired the mosquito for its persistence, recording in *Walden*:

> I was as much affected by the faint hum of a mosquito making its invisible and unimaginable tour through my apartment at earlier dawn, when I was sitting with the door and windows open, as I could be by any trumpet that ever sang of fame. It was Homer's requiem; itself an Iliad and Odyssey in the air, singing its own wrath and wanderings. (95)

This passage by Thoreau is a direct reference to a line in the *Iliad*, in which Homer wrote with admiration for "the persistent daring of the mosquito who, though it is driven hard away from a man's skin, even so, for the taste of human blood, persists in biting him" (*Walden: Annotated*, 87).

Thoreau's (and Homer's) perspective is probably not shared by many people today, who may recall evening barbeques ended prematurely by the onslaught of biting insects and sleepless nights caused by the whining sound of a mosquito approaching an exposed ear. But keeping mosquitoes' persistence in mind is important as we consider the effects that climate change might have on their life cycle in New England, where their presence is currently kept somewhat in check by temperature.

Weather also affects the daily abundance of mosquitoes. Mosquitoes are most abundant on calm, overcast, and humid days; aside from being slow, they are also fairly weak fliers and are easily blown about by the breezes. Thoreau was aware of how weather affected mosquitoes, writing after a rainy week on August 7, 1853, "Methinks the mosquitoes are not a very serious evil till the somewhat cool muggy dog-day nights, such as we have had of late." Thoreau is again making a classical reference, as the Romans believed that hot summer weather was caused by Sirius, the Dog Star, the brightest star in the night sky during summer in the Northern Hemisphere.

In New England, mosquito abundance varies greatly from year to year. People who spend time in the woods know that a wet spring and summer, when abundant snowmelt and rainfall fills ponds and swamps, will lead to large mosquito populations. If you've ever hiked through a forested landscape, you know that the density of mosquitoes changes from a few isolated individuals in dry woods to swarms of hundreds, or even thousands, upon approaching a swampy section of forest. No doubt you picked up your pace as you tried to leave the vicinity of the swamp as fast as possible. A rapid pace of walking does help to leave behind the cloud of mosquitoes, as they are not fast fliers. Also, female mosquitoes locate warm-blooded animals by homing in on the carbon dioxide and heat they emit, and they have trouble getting a fix on fast-moving animals.

One rainy May day fully three decades ago, I was carrying out fieldwork on lady's slipper orchids when I was unexpectedly engulfed by a cloud of hundreds of mosquitoes. I had on a short-sleeved shirt and no mosquito repellent. But I decided to ignore the mosquitoes and finish my task. After an hour I had over a hundred bites on my arms, face, and neck. Years earlier, when I was an undergraduate, I was hiking alone in the White Mountains of New Hampshire. It was dusk when I arrived at my campsite near a wet, woodsy area. As soon as I stopped, mosquitoes swarmed around me in the hundreds and fiercely hungry. My one thought was to build a fire as fast as possible and heap it up with damp wood and leaves to create the smoke that would drive them away. I did this as calmly as possible, as I was on the verge of being driven crazy by the buzzing in my ears and pinpricks on my skin. Even now, forty years later, I can see my hands covered with dozens of mosquitoes as I struck the matches to light the campfire.

Male as well as female mosquitoes actually rely on plant nectar, not blood, for energy; females alone dine on blood, and they store it in a separate stomach, or crop. The female mosquito uses the protein from her blood meal to form her eggs. Mosquito species feed on the blood of different vertebrate blood donors: some specialize on reptiles, other on birds, still others on mammals. Of the 3,500 species thus far described by scientists, only a small percentage are known to take a meal of human blood.

Mosquitoes that bite people are most active at dawn and dusk, as these are times when the air is humid, and the air relatively calm for flying. Most people living in metropolitan areas will be more likely to encounter the evening attacks of mosquitoes due to modern habits of getting up well after dawn. Campers are more likely to see the dawn raids of the mosquitoes, such as Thoreau did while on a summer trip to northern New Hampshire: "The mosquitoes troubled us in the evening and just before dawn, but not seriously in the middle of the night. This, I find, is the way with them generally" (*Journal*, July 6, 1858).

Mosquitoes and Climate Change Research

Due to the importance of spring warming and autumn frosts as well as rain and wind in defining the flight times and abundance of mosquitoes, I thought that mosquitoes in New England could be an excellent subject for climate change research. I began to formulate our research questions: How will mosquito numbers and behavior be affected by changing temperatures and rainfall patterns? Have changes already begun, and, if so, what are they? Given the changes we'd found in plants, birds, and butterflies, I thought it highly likely that a mosquito species that relied upon temperature to free it from icy ponds would likewise show alterations in behavior, abundance, and life cycle. However, I had no idea where to find such information. Where do you even start looking for information on mosquito abundance?

In the late winter of 2009, I was giving a public lecture on our climate change work as part of the Darwin Science Festival to a diverse group of students and professors at Salem State University just north of Boston. I concluded with a lament that, despite the abundance of bird and plant records that we had uncovered, we still had not discovered a comparable body of long-term records of insect phenology. And as always, I ended my lecture with an appeal for anyone in the audience to provide us with suggestions about how we could find more information.

After my talk, when most of the audience had departed, a distinguished-looking older man wearing an oddly formal black suit came up to talk with me. He introduced himself as Alfred DeMaria, a physician and director of the Massachusetts State Laboratory. He explained that his laboratory had been censusing mosquitoes for the past few decades. He believed that this information might be useful to us.

Dr. DeMaria later told me it was a revelation to hear me speak about the evidence of climate change in New England at Salem State University:

> With over fifty years of mosquito surveillance data collected by the Massachusetts Department of Public Health available, the opportunity to probe changes in mosquito number and emergence, by species, presented itself as a way of potentially assessing what role if any climate change was having on what we think we are seeing with the epidemiology of an important mosquito-borne disease in Massachusetts. It also offered the potential of looking at an insect indicator of the impact of climate change in New England and what effect it might have on public health.

I took his name and e-mail address, and offered to contact his staff during the next week to see what type of data they could share with us. The following week, I visited the state laboratory in Jamaica Plain with my graduate students Libby and Caroline, where we met Cynthia Stinson, the director of the mosquito project, and Matt Osborne, a mosquito researcher. It turns out that the state lab had enormous amounts of mosquito data because of concerns related to public health.

Mosquitoes and Eastern Equine Encephalitis

It's well known that mosquitoes spread diseases such as malaria, West Nile fever, and dengue fever. When a female mosquito feeds on an infected host, she takes microscopic pathogens (such as bacteria, protozoans, and viruses) as well as blood into her body. While the mosquito is now carrying the pathogen, the infection may do her no harm; she's merely a carrier, or vector. The disease-causing microbes circulate in her body, and many wind up in her salivary glands. When she bites another animal — such as a bird, rabbit, or human — the infected mosquito injects some of her saliva into the wound to prevent the blood from coagulating and to keep the blood flowing. Her saliva carries the microbes, which now circulate in the blood of their new host. Over the following hours and days, they multiply, starting a new infection in a previously uninfected animal or person.

Massachusetts has one of the oldest and most comprehensive mosquito-monitoring programs in the United States. The reason is both simple and tragic. Massachusetts is at the epicenter of repeated outbreaks of a deadly viral disease spread by mosquitoes, Eastern equine encephalitis, often abbreviated EEE and colloquially referred to as Triple E. The name of the disease refers to the fact that it is found in the eastern United States, the first victims observed were horses (*equine* refers to their Latin name), and the virus causes a swelling of the lining of the brain, known as encephalitis. The first record of this disease dates to the wet summer of 1831, when horses in Massachusetts were reported to be dying of "brain fever," a catch-all term that includes fever, excitability, and convulsions.

Another epidemic in August 1938 resulted in the death of over 300 horses and caused concern among the public that the disease might also affect people. At first, public health officials reassured the populace that equine encephalitis could not spread to people. Unfortunately, this proved untrue: 35 people, including 24 children, eventually came down with encephalitis — 25 of whom died, an extremely high rate of mortality. Tragically, most of the mortalities were young children. Typical symptoms of EEE in people are an altered mental state, including drowsiness, stupor, or even coma, convulsions, and stiff neck. These symptoms became permanent in most surviving patients, along with paralysis of limbs and reduced mental capacity.

There was another epidemic in 1955–56, with 85 horses being affected. . Of the 16 people who contracted the disease, 9 died. Less severe outbreaks also occurred in 1970–75, 1982–84, the early to mid-1990s, and 2004–6. Massachusetts was predicted to be at risk for major outbreaks in 2010, 2011, and 2012, due to the abundance of mosquitoes and the presence of the virus in sample collections of mosquitoes.

A peculiar feature of EEE is that is it occurs sporadically, with gaps of many years or even a decade or more with no recorded outbreaks. It is not known why there are long periods during which the virus is present in Massachusetts but appears not to cross over into horses or people.

Also, there seemed to be intervals of fifteen years or more between past outbreaks, but now the time between outbreaks seems to be only six years or even less. Are we just getting better at detecting cases of EEE, or is the number of years between outbreaks really getting shorter — and if so, why? And what can be done to stop the disease?

The state laboratory has been investigating EEE for the past fifty-three years. And they have learned a great deal already. Of key importance is that these outbreaks of EEE are all concentrated in southeastern Massachusetts in areas with extensive swamps of red maple and white cedar. These are mainly the towns south of Boston and northwest of the Cape Cod Canal, including Plymouth, where the Pilgrims landed over 350 years ago. The proximity of these outbreaks to areas with a lot of standing water, coupled with the fact that the disease mostly occurs in late summer, means that mosquitoes are strongly implicated as the carriers of the disease.

Work by state researchers revealed that a common mosquito of Massachusetts' swamps, *Culiseta melanura*, is the main carrier of the virus that causes EEE. *Culiseta* is a medium to large mosquito, with a distinctive downward-curved proboscis. Its larval stage lives in the acidic freshwater swamps of southeastern Massachusetts. Adult mosquitoes sometimes test positive for the presence of the EEE virus, whereas the dozen or so other mosquito species living in this region rarely or never test positive. So *Culiseta* is a key vector in disease transmission.

Catching Mosquitoes

Cynthia Stinson, Matt Osborne, and others at the state lab have carried out an extensive program to trap mammals and birds in the swamps of southeastern Massachusetts and collect their blood. Testing has demonstrated that common forest songbirds also serve as a reservoir for EEE in years when there is no disease in horses and people. This makes sense, as birds are the preferred blood source for *Culiseta* mosquitoes. But *Culiseta* doesn't seem to feed on humans or horses; that's done by two other mosquitoes also present in Massachusetts swamps, *Aedes* and *Culex*. Where the epidemiology becomes complicated is figuring out how the EEE virus gets from songbirds and *Culiseta* to horses or people. Even though *Aedes* and *Culex* mosquitoes are suspected of transferring viruses from birds to people and horses, these mosquitoes rarely test positive for the virus. It's a complicated problem — but my students and I were less interested in the puzzle of EEE transmission and more in whether and how *Culiseta* had changed its ways over time due to a warming climate.

Following the epidemic of 1955 and 1956, the state of Massachusetts began the trapping of mosquitoes at swamps throughout southeastern Massachusetts to determine seasonal patterns of abundance and to screen mosquitoes for the presence of the EEE virus. Over the past half century, the program has trapped hundreds of thousands of mosquitoes using many different methods. Fortunately from my lab's perspective, they have also consistently sampled ten sites in large swamps using the same methods throughout this span of time. At each site, mosquitoes are trapped once a week from late May until late September or early October. At each site, usually two traps are hung on trees on one day and picked up the following morning. A light in the trap attracts mosquitoes, and a fan draws the mosquitoes into a mesh bag, preventing them from escaping. This type of trap is particularly effective for trapping *Culiseta* mosquitoes that are attracted to light, but is less effective at trapping the *Aedes* and *Culex* mosquitoes that bite people; for these mosquitoes, traps baited with carbon dioxide are more effective. Mosquitoes caught in the traps are brought back to the state lab and tested for the presence of the EEE virus.

In the summer of 2009, Matt Osborne and I, along with my twelve-year-old son Jasper, visited one of these sites. After heading south from the laboratory for about one hour, we left the paved road and drove for a mile down a very rough dirt road into the middle of a huge swamp. This was not the type of place people normally visit intentionally; just off the road on both sides, the place was filled with standing water and shrubs, and was extremely dark on this overcast day.

Before leaving the truck, each of us put on a special pullover jacket that included fine-meshed netting to cover our head and neck. We were also wearing long pants, socks, and gloves, with no bare skin uncovered. Walking around the swamp for an hour, knee-deep in smelly water and ankle-deep in stinking muck, with swarms of deerflies crashing into the netting and clouds of mosquitoes looking for any tiny patch of exposed skin to bite might not be everyone's idea of fun. But it's what you need to do to understand the outbreaks of EEE.

Mosquitoes Connect Birds and People

According to reports written by scientists from the state laboratory, the life cycle of the *Culiseta* mosquito makes it and the EEE virus particularly sensitive to both short-term weather patterns and long-term changes in the climate. The larvae of *Culiseta* overwinter in the cold water of swamps, with their development directly controlled by temperature. At water temperatures of 50 degrees Fahrenheit and below, which occur throughout the winter, *Culiseta* larvae stop developing. When swamp water temperatures warm to 50 to 60 degrees, the larvae begin to grow slowly, and at temperatures above 60 degrees, the larvae grow more rapidly; at optimal temperatures above 60 degrees, the larvae take three months to develop. The result is that *Culiseta* has two distinct pulses of adults during the growing season, a life-cycle pattern called "bivoltine" by entomologists. One group of adults emerges from the swamp in May, feeds on birds, lays eggs in the swamp water, and dies. This second batch of larvae take three months to develop in the warm summer water, and a second generation of adults emerges in August, feeds on birds, and lays eggs again; the resulting larvae overwinter and emerge in May, starting the cycle all over again.

This bivoltine life cycle has major consequences for the buildup of virus populations in birds. In the spring, *Culiseta* feeds mostly on the adult resident birds, such as the cardinal, tufted titmouse, chickadee, and robins, and the newly arrived migrants, such as the wood thrush and the catbird. These adult birds are predominantly healthy, uninfected individuals who have previously been exposed to the virus and are resistant to EEE. As a consequence, the first generation of *Culiseta* adults that emerges in springtime does not generally test positive for the virus. However, when the second generation of *Culiseta* adults hatches in late July and August, most of the bird populations consist of juvenile birds with no prior exposure to the virus. When summer mosquitoes bite adult birds with some residual virus in their blood, and then bite juvenile birds, the virus population has a chance to rapidly expand in the bodies of the juvenile birds that have no antibodies to protect them from infection. Many of these infected young birds, perhaps robins in particular, will build up large concentrations of viruses in their bodies. These first infected juvenile birds, especially if they are sick and less able to fend off the attacks of mosquitoes, become a principal source of EEE infections that mosquitoes will then transfer to additional juvenile birds throughout the area. Over the course of days and weeks, millions of mosquitoes attacking thousands of birds will cause of the buildup of billions of virus particles within the bird population.

Based on this scenario, we can see that the largest buildup of the EEE virus in these swamps will take place in years with mild temperatures in winter and spring and abundant spring rains so that the swamps and other bodies of water are filled and offer plenty of water surface and volume for mosquito larvae. These factors ensure that the first generation of mosquitoes produces more offspring that have a better likelihood of survival — and a larger population of mosquitoes in the second generation means that birds, especially that spring's fledglings, are bitten multiple times. Similarly, mild winter and spring temperatures are good for nesting birds, and juvenile birds must also be at high density to facilitate the easy transfer of virus from bird to bird by the *Culiseta* mosquitoes. Both factors must align in order for there to be an EEE outbreak. An additional factor that might be contributing to the declining number of years between outbreaks is the growing populations of American robins that overwinter in Massachusetts. In the past, robins always headed south for the winter. However, due to a warming climate and increasing populations of invasive plants with fleshy fruits, robins now increasingly spend the winter in eastern Massachusetts. Young robins are apparently readily bitten by mosquitoes and in the process serve as reservoirs for the virus, which are then transferred to new mosquitoes.

There is also some recent evidence that each new outbreak of EEE is caused by a different genetic variant of the virus. This suggests that each outbreak of the virus in bird populations is also related to the ability of a novel form of the virus to overcome the acquired immunity of birds. And the declining number of years between outbreaks could be due to a more rapid arrival of new variants of the virus from outside of the region or a more rapid mutation rate.

The missing piece of the puzzle is how the virus jumps from bird populations to horses and people. Because *Culiseta* feeds almost exclusively on birds, other mosquito species that feed on both birds and mammals, including horses and people, are considered to act as bridge vectors, spreading the disease more widely. These include genera such as *Aedes* and *Culex*, which readily attack people and are notable for spreading to humans such diseases as West Nile fever, yellow fever, and dengue fever. When one of these mosquitoes bites an infected fledgling bird and then a person, the disease can cross over into the human population. However, *Aedes* and *Culex* mosquitoes rarely test positive for EEE. So at this point, researchers remain uncertain if the relatively rare crossover of EEE to people and horses is being accomplished by the very rare occasions when infected bird-loving *Culiseta* mosquitoes bite people and horses, or alternatively by rare *Aedes* or *Culex* mosquitoes that are infected with the virus.

Decades of research on mosquito populations and past outbreaks have been used to develop predictions of the chance of an outbreak of EEE among people in southeastern Massachusetts in a given year. The most important predictors are large numbers of *Culiseta* larvae and adults in both the spring and summer broods, and a high percentage of adults infected with the virus in the summer brood. Large populations of other mosquito species, such as *Aedes* and *Culex*, that bite people late in the summer are also thought to contribute to outbreaks.

The value of horses as an early indicator of later outbreaks of human EEE cases declined after 1973 when a vaccine for horses became available. Virtually all of the approximately 25,000 horses in the region are now vaccinated, resulting in far fewer EEE cases involving horses; and those cases of EEE in horses now come at approximately the same time as the human cases. However, in 2010 the first EEE case of the year occurred in a horse, and this attracted great attention from the Massachusetts government.

At this point, my students and I have all of the mosquito data from the Massachusetts State Laboratory. We are deciding how best to analyze this mass of information to determine if climate change is altering the timing of mosquito populations and their rate of infection with the EEE virus. However, the bivoltine life cycle and the great week-to-week variation in mosquito abundance make the analysis extraordinarily complex and challenging. A further difficulty in interpreting the data is that two traps were not consistently used per site, which can strongly affect the numbers of mosquitoes caught. We are coming to the conclusion that we may need to expand our research team to include scientists who specialize in analyzing changes in mosquito abundance.

Spraying to Control Mosquitoes

Some scientists have suggested that with a warming world, many disease-causing mosquitoes will expand their range out of the tropics into subtropical and temperate areas. In this view, mosquitoes will closely track temperature, bringing tropical diseases, such as dengue fever and malaria, more widely into developed countries like the United States. However, more careful analysis suggests that the ranges of mosquito-borne diseases are shrinking and their severity is declining, at least in tropical countries, due to the increased use of insecticides, the draining of wetlands, the use of screens and mosquito netting in houses, and people moving from rural areas to cities.

Public health officials in Massachusetts are using the spraying of insecticides as a key tool in their efforts to reduce the risk of EEE. The other actions involve educating the public to take precautions to avoid being bitten by mosquitoes: avoiding outdoor activities at dawn and dusk when mosquitoes are most active; covering exposed skin with hats, long-sleeved shirts, and pants; and using mosquito repellents.

For the past fifty-seven years, the Massachusetts state government has been carrying out spraying programs to kill mosquitoes in the southeastern part of the state in years when there are numerous infected mosquitoes. Unfortunately, spraying carries with it risks of its own. The insecticide DDT was applied in 1955 and 1956, but when studies showed it had negative effects on wildlife, particularly hawks, eagles, and ospreys — not to mention potential long-term health effects on people — it was banned. During the outbreak in the 1970s, another insecticide with fewer side effects, Malathion, was substituted for DDT, with 1.7 million acres treated in 1973 and 82,000 acres in each of 1974 and 1975. More recently, the state of Massachusetts has used a chemical called Anvil, which is a synthetic version of a natural insecticide found in chrysanthemum plants. Although Anvil is thought to be less toxic than DDT and Malathion, and to linger for shorter periods in the environment (unlike DDT, which can remain in soil and groundwater for extended periods), there are still valid concerns about its potential for triggering allergic responses and disease.

A further difficulty with spraying is that it is unknown how effective it is at reducing the number of human cases of EEE. Spraying certainly kills many mosquitoes, but large numbers survive because the spray misses them due to erratic wind currents; because they are under leaves, trees, or some other object; or because they emerge from the water in the days after spraying. Since cases of human EEE are so rare and so erratic from year to year, it is difficult to determine how many cases, if any, spraying really prevents. Spraying took place in 2006, but five people still contracted EEE and two died. Spraying was done again in 2010, with one man getting EEE but then recovering. Does that mean that the spraying did not work, or that there would have been more human deaths without the spraying? We do not know. A very strong argument can be made for not spraying and putting more resources into informing the public about the disease and strategies for avoiding getting bitten by mosquitoes. Another argument against spraying is that it also kills a wide variety of insects, in addition to mosquitoes, and at least some of these insect species are ones that eat mosquitoes, including dragonflies. As a result, once the government begins to spray wetlands to reduce mosquito populations, spraying may need to happen almost every year as the natural predators of the insects have been removed from the ecosystem.

The Importance of Insects

Throughout our ten-year search for insect data, a number of larger developments have served to remind us how serious climate change could be for humanity. During our search, colony collapse disorder of honeybee hives reminded us that bees of all kinds, but particularly honeybees and bumblebees, are vital for pollinating crops. We wondered whether rising temperatures could have anything to do with that problem. Perhaps increasing temperatures were making bees more susceptible to the chemicals, parasites, or pathogens that are considered to be the leading suspects in this widespread environmental and agricultural calamity. Whether or not climate change turns out to be a factor in the current episodes of colony collapse disorder, a warmer, wetter world could expand the range of a number of pathogens that are known to target honeybees. Similarly, the expansion of the pathogenic chytrid fungus, which is wiping out frogs in much of the tropical world, is believed to be fueled in part by climate change.

Once we consider the mosquito as a prime candidate for the role of indicator species about insects and climate change, we're immediately confronted by the mosquitoes' role as a disease vector. At present, our mosquito data comes from research to track equine encephalitis, but malaria poses an even greater threat. Thought to have been eradicated in the United States over sixty years ago, malaria is the scourge of the tropical world, and it is not hard to imagine the disease becoming reestablished in the United States as Central and South American mosquito species expand their ranges northward. Other diseases, including West Nile virus, have already come to the United States from much farther afield. And even mosquito species from tropical Asia and Africa can hitchhike to the United States on airplanes and container ships, and establish themselves in a warming climate. Looking at aquatic insects leads us to realize that there are direct consequences to human health and food supplies when the climate changes and insect populations change in abundance.

The latest surveys indicate that most Americans accept the idea that the climate is getting warmer. Further, they agree that extreme heat waves and other unusual weather events are caused by climate change. However, this awareness of the effects of climate change has still not translated into significant political action. Americans may not see the threat to human health and well-being posed by the loss of pollinators, the disruption of freshwater food webs, and the expanding ranges of disease-carrying insects. To them I say, consider the monarch.

Many people my age remember when monarch butterflies flew south in the late summer by the hundreds and thousands; now, my youngest child is lucky to spot even a few migrating monarchs. And yet even one monarch sipping nectar from a pink milkweed flower in the summer can remind us of why we need to take climate change and other threats to the environment seriously. The threats to monarch butterflies from overharvesting of timber in Mexican rain forests, rising winter temperatures, and chemical pollution also have negative impacts on people. We need to protect nature for its own sake, as well as for our health, prosperity, and enjoyment. Do we want our children and grandchildren to grow up in a world where the only butterflies they see are in their picture books?

Green frog in Concord; photo by Tim Laman.

*In almost all climes, the tortoise and the frog are among the
precursors and heralds of this season [spring], and birds
fly with song and glancing plumage, and plants spring and bloom,
and winds blow, to correct this slight oscillation
of the poles and preserve the equilibrium of Nature.*

THOREAU, *WALDEN*

12. The Frog Chorus

THOREAU HAS A GREAT DEAL to say in his writings about birds, plants, and insects; he says somewhat less about amphibians. This is disappointing for our research because the life cycles of amphibians are also seasonally determined and may be affected by climate change. Many amphibians — such as the eastern newt, the American toad, and the spotted salamander — lay their eggs in water in the early spring, thus linking their breeding to the timing of ice-out in bodies of water. Male frogs and toads will arrive at water bodies in the early spring, just after the ice has melted, and start calling for females. In chapter 8, you will recall that Betty Anderson recorded the first spring mating call for the males of spring peepers, the American toad, and the wood frog. We found that the calling of spring peepers was sensitive to temperature, with choruses starting around dusk on the first warm days of spring. A more comprehensive, long-term study of spring activity could, in theory, provide the same sort of evidence for the effects of climate change on amphibians that we've found among insects, plants, and birds.

I say "in theory" because reports of frogs, toads, and salamanders vanishing from aquatic ecosystems have been in the news off and on for the better part of three decades, although the reports have intensified since the Global Amphibian Assessment in 2004. This was a comprehensive global survey of the conservation status of all known amphibians, carried out under the auspices of the International Union for Conservation of Nature (IUCN), a consortium of the world's conservation organizations. The most recent survey of amphibians indicates that fully 36 percent of the world's amphibian species are in danger of going extinct. In the United States, the percentage of endangered species could be as high as 42 percent.

Given that mating and egg laying of many amphibians are tied to the spring temperatures in much the same way that birds' are, it seems probable that if you examine the ways that amphibians' life cycles and populations are changing, you'll likely find evidence that global warming is affecting them. But of course it's not that simple. Many amphibians in New England aren't well studied simply because they're so difficult to find. They have secretive habits, are only active at night, and often have erratic, highly variable reproductive cycles — with multitudes of individuals mating in one year and virtually none the next. Such circumstances make obtaining solid, consistent evidence about their population changes challenging for experienced field herpetologists and all but impossible for my students and me.

Trouble for Turtles

The challenge of working with aquatic animals was brought home to me when I spent a few hours working with wildlife ecologist Bryan Windmiller and his crew searching for the Blanding's turtle, an endangered reptile, at Great Meadows National Wildlife Refuge. Because of its protected status, there is a great deal of effort being placed on determining where and how long these turtles live and how many there are. On this day, we were trying to locate a particular individual that had been fitted with a radio transmitter on its back. This should have been easy, as we had a directional antenna that could locate the turtle to within a few feet. Unfortunately, as we wandered around in the knee-deep mud of the shallow marsh, the turtle was not cooperating. It remained motionless and buried in the mud. Even though we were within a foot or two of the turtle, we could not find it. The sun was beating down on me, the sweat was running down my face, and my shoes and pants were covered with slimy mud. Only after half an hour of grubbing around did someone finally grab the Blanding's turtle. The object of our attention was a mud-covered eight-inch-long turtle with a rounded shell and a yellow throat. At times like this, I felt grateful for being a botanist.

Considering that we had so much trouble finding this turtle, it seems that the Blanding's turtle should be able to avoid the problems of the modern world. However, this is not the case, as described by Jon Regosin, the Massachusetts state officer in charge of protecting endangered species:

> Blanding's turtles and roads just don't mix. The Blanding's turtle has evolved a reproductive strategy, over millennia, which offsets very low nesting success and juvenile survivorship with very long-lived adults. In their teens when they reach reproductive maturity, these turtles need to live a long time, on average, in order to successfully reproduce. Therefore, even small increases in adult road mortality can spell disaster for Blanding's turtles and other vulnerable turtle species. When you combine these local threats with the uncertainty that comes with climate change, you begin to see the full picture of the pressures that our rarest and most imperiled species are facing.

Regosin mentions climate change in particular because rising temperatures or changing rainfall patterns could dry out many of the marshes where these endangered turtles live.

Climate change will not just be affecting Blanding's turtles in the future. Windmiller has already observed that whereas these turtles used to lay their eggs in mid-September at Great Meadows, they are now laying eggs two to three weeks earlier in late August and early September. An even more insidious effect of a warming climate is its potential impacts on the sex ratio of turtles. Temperature affects the sex of turtles and other reptiles during development in the egg. In a few recent years, most or all of the Blanding's turtle hatchlings have been female. A major negative consequence of having an unbalanced sex ratio in the turtle population due to a warming climate is that females will be unable to find mates, and the population will decline to extinction.

Too Dry or Too Wet: Changing Rainfall and Amphibian Life Cycles

And consider the impact of climate change on salamanders, slender amphibians with four legs and long tails that mostly live in moist habitats, under rocks and logs, and in freshwater wetlands. Salamanders should be among the most sensitive amphibians to climate change. In a warming world, the survival of salamanders and other moisture-dependent amphibians will be threatened as terrestrial habitats dry out during warmer summers. Salamanders and frogs take in oxygen, or "breathe," through their moist skin. Just a few days of above-average temperatures will reduce the humidity of the forest floor to such an extent that moisture-loving salamanders and frogs will not be able to forage at night and so may starve to death. If hunger drives them to forage even on dry nights, they may lose too much water and dry out and die or suffocate.

Some salamander species, in particular mole salamanders (so-called because they live underground), breed in pools of water, most often ephemeral water bodies present only in the springtime and summer, known as vernal pools. Such pools are shallow depressions that fill with water as the snow melts and continue to fill with spring rain. The water level in these pools gradually declines in the late spring and summer, and pools typically completely dry out in the heat of the summer. Because they dry out during the summer, these habitats cannot support populations of fish. And because there are no fish, vernal pools often teem with an abundance of aquatic species unable to reproduce in ponds and lakes where fish prey on them or their young. In these fish-free pools, they can complete their life cycle quickly, safe from predation.

In Concord these vernal pool species include the juvenile stages of two species of mole salamanders (the yellow-spotted salamander and the blue-spotted salamander), wood frogs, and American toads, as well as short-lived fairy shrimp. Aquatic insects — such as diving beetles, water striders, and giant water bugs — are often abundant in vernal pools, though they are also found in other wetland habitats.

ON MARCH 13, 1855, Thoreau recorded in his journal seeing many tadpoles — "medium or full size" — under a thin film of ice in the deep ditches in Hubbard's meadow: "I am surprised to see, not only many pollywogs [tadpoles] through the thin ice of the warm ditches, but in still warmer stagnant unfrozen holes in this meadow half a dozen small frogs."

In his journals, Thoreau frequently writes of spotting tadpoles in ditches, springs, and pools during his walks — often very early in the spring, when the puddles and pools are still glazed with ice. The greatest threat to many amphibians might be the loss of vernal pools, their crucial breeding habitat, as climate change brings warmer and drier conditions. All vernal pools dry out — that is, after all, what keeps them free of fish and makes them great places for amphibians to breed. But the timing of drying out is what matters, and for amphibians it's a matter of life-and-death.

Just when a particular pool dries out in the summer will depend on its depth (how much water there is), how hot it is each month (how quickly water evaporates), and the amount of rain that falls (how much water is added back into the system). In dry, hot years, the pools will dry out in the spring or early summer, but in wet, cold years, the pool might have water continuing until September or October. If the climate gets just a bit warmer, increasing evaporation of the water in the vernal pool, or if it rains just a bit less, then the vernal pond may dry out too early in the spring for some amphibian and insect species to finish their life cycle and metamorphose into adults. As a result, they will decline in abundance and may go locally extinct. Species with a shorter life cycle may do better and increase in abundance. The key point is that changing temperatures alter the environment, in effect creating winners and losers of particular species.

But just how climate change will alter patterns of precipitation is hard to predict. Global warming can mean *increased* precipitation in some places, or precipitation falling as rain instead of snow, in theory creating more vernal pools and expanding the habitat available to salamanders. But it doesn't always work that way. The spring of 2009 was one such unusually wet spring, so you'd think it would've been a banner year for Walden's amphibians, but it was a disaster for salamanders, as I will describe.

Thoreau noted the ability of vernal pools to cycle between wetlands and dry land: "I was accustomed to fish from a boat in a secluded cove in the woods, fifteen rods from the only shore . . . , which place was long since converted into a meadow" (*Journal*, March 13, 1855). Thoreau was likely referring to Wyman's Meadow, a large vernal pool next to Walden Pond, separated from it by a sandbar. Hikers following the trail around Walden Pond cross this sandbar just before they reach the former site of Thoreau's cabin. Wyman's Meadow is roughly circular and about fifty yards across. It is fringed on three sides by trees and is open on the fourth side where it adjoins Walden Pond. In typical years, the meadow fills with water in the winter and late spring to a depth of around three feet, forming a large vernal pool. It is perfect amphibian breeding habitat, and in the spring you can find abundant gelatinous egg masses, including mole salamander eggs, giving way to thousands of tadpoles, especially those of the American toad, and tiny juvenile aquatic mole salamanders a few weeks and months later. By summer the pool dries out, and the meadow becomes a carpet of flowers, such as pink-flowered meadow beauties, with their bright yellow stamens, and sprays of pink-and-white knotweed flowers. In dry years only a small pool of water remains in the middle of the meadow, and there is probably not enough volume of water or enough weeks for the amphibians to complete their life cycle.

In the rainy years that Thoreau mentions, similar to what we experienced in 2009, Walden Pond rises above the sandbar and takes over the meadow. In such years, fish from Walden Pond also cross the sandbar and enter the flooded meadow. Fishermen in turn follow along and catch fish in this temporary extension of the pond. But for the mole salamanders, wood frogs, and toads that breed in Wyman's Meadow, these extra-wet years are a disaster, as their young fall prey to hungry fish.

Salamanders in Eastern Massachusetts

In March 1854 Thoreau caught a salamander in the vernal pool ("spring hole") behind Hubbard's spring. He described it in his journal as

> . . . 3¼ inch long, tail alone 1½+, a dozen or more marks as of ribs on each side. It was lying on the mud in water as if basking. Under microscope all above very finely sprinkled black and light brown — hard to tell which the ground. I still have not identified it. Somewhat like S. dorsalis [the northern zigzag salamander], but not granulated nor ablated with vermillion spots, being uniformly dark above except to a microscope, beneath bluish slate; beneath & sides of the tail dull golden. (Journal, March 22, 1854)

Based on Thoreau's description, he seems to have found a red-backed salamander — though the leadback form lacking the red stripe that is common, but not universal, to the species. It clearly is neither of the two mole salamanders common to Concord: the blue-spotted salamander and the yellow-spotted salamander. These species breed in vernal pools and have aquatic larvae. Both species are lizard-like in shape, with an elongated tail and the soft, glistening skin of a frog. The two species are dark in body color, tending toward shades of black, with hints of dark blue, dark gray, and dark brown. The yellow-spotted salamander, the larger of the two, measures four to seven inches long, with ten or more large yellow spots running in lines down each side of its body. The blue-spotted salamander is somewhat shorter, four to five inches, flecked with bright blue patches all over its dark body.

Though it's hard to say for certain just what kind of salamander Thoreau was looking at, it wouldn't be surprising to learn that there are species other than the red-backed salamander or the blue- and yellow-spotted salamanders in Concord. Salamanders are notoriously cryptic — so well camouflaged that they are hard to detect against the background of leaf litter. Red-backed salamanders can be found if you know where to look, such as under flat rocks and decaying logs, but the two spotted salamander species are almost never seen, even by keen naturalists, due to their small size and secretive lifestyle. Adult spotted salamanders live most of their lives deep beneath the leaf litter and under logs and rocks in damp forests. They are also known as mole salamanders because of this tendency to burrow in the ground. Belowground they feed on insects, worms, slugs, spiders, and other small invertebrates that they find in the soil layer. On damp nights, spotted salamanders may leave their burrows to hunt for food on the surface. And, of course, if the forest becomes drier due to warmer nights, they are less likely to come aboveground, where they will be vulnerable to drying out.

The one time of year when spotted salamanders leave their earthy homes is early April, when the snow has melted from the ground and there is a relatively warm rain. At this time of year, always at night and always under the wet weather conditions they need to survive, spotted salamanders will leave the forest and migrate hundreds or even thousands of yards to nearby vernal pools, often the one that the individual grew up in. It is on such nights that spotted salamanders can be seen, sometimes in surprisingly large numbers, along known migration routes, and especially when these routes cross paths and roads.

Once enough males and females have arrived in the pond, they begin to swarm together, after which males and females pair off and move away from the swarm. The male then produces a sperm-filled packet (spermatophore) that he attaches to the bottom of the pond or dead leaves at the pond edge. The male guides the female over to his sperm package. She nestles on top of the packet, guiding it into a vent on the bottom of her body, where the sperm fertilize her eggs. She will deposit a mass of gelatinous eggs in the pond a day or two later, which will hatch into tiny aquatic salamander larvae with external gills.

A few months later, or when the pond dries out in summer, the juvenile salamanders undergo metamorphosis into the adult form and leave the pond for the adjacent forest habitat, where they will spend the rest of their lives. The salamanders have some flexibility in the rate of development and can metamorphose earlier if water levels are low and food is running out. However, if the pond dries out too early, especially if springtime temperatures are warmer than usual, the young salamanders will not be large enough or mature enough to make this journey, and most will die. Those species that are flexible and can time metamorphosis based on current conditions may outcompete less flexible species.

Salamanders in the Hammond Woods

During more than fifty years of exploring the Hammond Woods in Newton, I have seen thousands of the common red-backed salamanders. For the first forty-five years of my life, I thought that they were the only salamanders in these woods. As a boy, I would frequently find these slender, two- to four-inch-long, reddish or gray salamanders when I turned over rocks and logs. I paid repeated visits to one flat rock, about two feet across and eight inches thick. When I lifted it up on one side, nine times out of ten I would find one or more red-backed salamanders. Sometimes there were as many as six salamanders under this one rock, with both the red form and the dark gray form, and often a mix of adults and the smaller juveniles. Years later, lifting this rock to look for salamanders became a favorite feature of the walks with our three children, to the point that we even gave it a name: Salamander Rock.

However, despite going for thousands of walks in the Hammond Woods and turning over many thousands of logs and rocks (and always replacing them back to their original position), I have only seen secretive adult spotted salamanders on two occasions, which suggested to me that they are quite rare. The first time was in 1995, when I pulled up Salamander Rock during a walk and, to my amazement, found that instead of the usual diminutive red-backed salamanders, there was a yellow-spotted salamander underneath. It looked enormous to me at the time, as it was so much thicker-bodied and longer than the slender red-backed salamander I was used to seeing. After studying the creature for a few minutes in stunned silence, I carefully replaced the rock so as not to disturb it. When I returned the next day, it was gone, and I never saw another one in that spot.

This sighting told me that there *were* spotted salamanders in the Hammond Woods, despite the fact that I had seen my first one only after forty years of looking. I next saw a spotted salamander thirteen years later, in mid-April 2008, as I was crossing the train tracks that run through a section of the woods. I noticed a dead yellow-spotted salamander in the groove of the rail, where it had probably been crushed by a passing train. The salamander had probably been trying to cross the train tracks to breed in an old cranberry bog, which was now a large vernal pool.

The fact that I've seen so few of them doesn't mean they're particularly rare. Even in the nineteenth century, Henry David Thoreau regarded finding spotted salamanders in Concord to be such a rare occurrence that he noted this unusual event made while gardening: ". . . from under a rotten stump my hoe turned up a sluggish portentous and outlandish spotted salamander, a trace of Egypt and the Nile, yet our contemporary" (*Walden*, 173).

I'd like to believe that they're relatively common, just hard to spot, and I've reason to believe this could be the case. In the spring of 2009, the exceptionally wet spring that I have already mentioned, the vernal pools of the Hammond Woods were filled to the brim. My favorite vernal pool is called Bare Pond, because it is a pond all during the winter, when local people ice-skate on it, but is "bare," meaning without water, in the summer. This pond is about forty by twenty yards and is fringed with red maple trees above and blueberry bushes below. In the spring I enjoy lying on rocks at the edge of the pond and studying the diversity of small insects, red water mites, and assorted crustaceans moving through the water between the dead leaves on the bottom and the pond surface. By July 2009, all of the pond creatures were increasingly concentrated in just a few remaining pools of water still left in the lowest parts of the pond. As I was observing this aquatic zoo, I noticed a salamander resting a few inches below the surface on a dead leaf on the pond bottom. At first I thought that it must be a red-backed salamander, based on its size of about two inches long. But then I realized that the red-backed salamander does not live in the water as either a juvenile or adult, so my initial assumption could not be correct. Thinking it over carefully, I realized that this must be the larval stage of the yellow-spotted salamander, something that I had never seen before at Bare Pond or in any vernal pond in these woods.

When my twelve-year-old son Jasper and I returned the next day with a net, we caught anywhere from one to a half dozen of these aquatic salamanders every time we dipped our net into the leaf litter and mud at the bottom of these pools. The salamanders were greenish brown in color and had slightly enlarged heads, but none of them had any spots, so we could not tell whether they were the juveniles of blue- or yellow-spotted salamanders. However, they were most likely yellow-spotted salamanders, as I had already seen the adults in these woods. And since they did not have their external gills, it was clear that they had matured enough to leave the pond. Based on the number of salamanders we were catching and releasing in every sweep of the net, there must have been many hundreds or even thousands of individuals that would soon be leaving for the nearby forest. Yet it was sobering to realize that they would never be seen by any person living in the area, except the most dedicated and persistent naturalist. No one would suspect that such a creature lived in these woods.

Salamanders at the Golf Course

There is one place in Concord where patient naturalists can sometimes see spotted salamanders with a greater likelihood of success. On the golf course of a suburban country club, only a short distance from Route 2 and the Emerson Hospital, a low hill slopes gently down to a vernal pond about forty feet across. Numerous yellow- and blue-spotted salamanders make their year-round homes in this varied landscape. On the first relatively warm, rainy nights in April, these salamanders migrate down the hill to breed in the pond. Unfortunately for the salamanders, the golf course has an access road directly across the migration route and a parking lot at the base of the hill. The salamanders must descend from the hill, cross the parking lot, and navigate the two-lane access road in order to reach the safety of the pond where they breed. Of course, they must cross this dangerous terrain again on their way back up the hill after they breed. To add to the challenge, curbs bordering the parking lot create vertical walls that are insurmountable for many of the salamanders.

Both to observe the salamanders and to help them along, our family has made it a point to visit the golf course pond each spring on the first warm, rainy night of April. The number of salamanders can vary from dozens of both species to a handful of just one species, and sometimes there are simply none.

On our first family outing to the golf course to help the salamanders, on an April evening seven years ago, we were joined by Jon Regosin and his family. Regosin works for the Massachusetts Division of Fisheries and Wildlife, enforcing the regulations that protect endangered species in the state. Much of his job involves working with real estate developers to make sure that new real estate developments do not harm protected species, or at least inflict the minimum possible damage. Regosin's work often involves redirecting construction sites to parts of a property where there are no such protected species, in the process creating special conservation zones where endangered species can live in peace. Since many of the endangered species in Massachusetts are wetlands species, protection of endangered species often means protecting wetlands and adjacent habitat that support wildlife populations.

As befits someone with such an important and sensitive job, Regosin speaks slowly and chooses his words carefully, so he will not be misinterpreted. But despite this calm exterior, Regosin can become animated when discussing amphibians, turtles, and other reptile species. He is passionate about these animals, having studied them for his doctoral dissertation. If I ever call him or e-mail him with a salamander question, he responds within hours and at length. For other questions, he can take quite a while to respond.

Our rainy night of salamander watching fell on a school night, and in a departure from our usual routine, we piled our kids into a minivan, along with flashlights and umbrellas. Jasper was nine years old, and Regosin's children were around the same age; the kids were excited about seeing the migrating spotted salamanders, and the grown-ups encouraged the kids by emphasizing how the salamanders needed their help to cross the golf course road. When we arrived, we had the parking lot to ourselves; no one plays golf in the dark and rain! We got out and began training the beams of flashlights around the parking lot. Once we saw our first migrating spotted salamander, the children became excited. Every few minutes, another salamander would emerge from the forest edge and start to cross the parking lot. Over the course of an hour, we encountered several dozen of them. The salamanders moved ponderously, making them both easy to find and easy to pick up. They were crawling along so slowly that to use the word "catch" would be inaccurate.

In the dimly lit parking lot, our two families were finding mostly yellow-spotted salamanders and only a few of the blue-spotted salamanders. The females were distinctly thicker around the middle as they were carrying a load of eggs. As we found each new salamander, Jasper and Regosin's children would gently gather up the salamanders in cupped hands and carry them across the parking lot and access road. Safely across, each salamander was released on the grassy slope leading down to the pond. This was a night to remember; a night for salamanders, and not one of the hundreds of typical nights of homework, computers, television, and reading.

Most salamanders that night were marching steadily off the hill and heading across the parking lot to the pond. But a number of the salamanders seemed disoriented by the parking lot and in particular by the curbs, and began to head off at right angles to the pond. One of the children in the group would then have the thrill of picking up the lost salamander and carrying it to the slope above the pond, down which it readily began to descend.

Spotted salamander populations have declined and gone locally extinct at many sites in Massachusetts, and the blue-spotted salamander is listed as a Species of Special Concern. My impression is that this road and parking lot at the country club must be having some negative effects on the homing movement of salamanders, but at this point, I really do not know for certain. Perhaps the area currently has the maximum possible number of spotted salamanders or the numbers of salamanders are declining over time. Only long-term studies could reveal the trends, but this information is not available. Gathering such long-term data is difficult, as salamander numbers vary extremely from year to year. In some years, large numbers of salamanders may migrate from their forest habitat to breed in the nearby pools, and in other years relatively few spotted salamanders may breed. Also, it would not be easy to obtain funding for a species that is not officially listed as endangered. The challenges of censusing spotted salamanders are made even more difficult because individual salamanders appear to be able to live for many years, possibly even a decade or more, so they can wait out years that seem unfavorable and just breed in the best years. A good breeding year for a salamander might be a preceded by a year with plenty of insect food and rainy moist nights, followed by a winter and spring with the abundant snow and rain needed to fill the vernal pools to full capacity.

There may be hundreds or thousands of spotted salamanders in the Concord woods, yet they will never be seen by people walking in the area or even by serious naturalists who go looking for them. Even if a biologist like me seeks them out during rainy nights in early spring, the salamanders may simply not migrate that year, or they may be migrating earlier or later in the spring, or they may be migrating later at night when no one is around to watch. It is very hard to study spotted salamanders. So it's equally hard to know whether they're responding to climate change at this time — and if so, *how* they're responding.

The Future of Amphibians in Concord's Woods

As much as the singing of birds, the calls of frogs, from the spring peeper to the deep-throated bullfrog, signal the arrival of spring. But not everyone likes the frog chorus, as Thoreau noted in his journal:

> *Mr. Samuel Hoar tells me that about 48 years ago . . . there used to be a great many bull frogs in the mill-pond which by their [trumpeting] in the night disturbed the apprentices of a Mr. Joshua Jones who built & lived in the brick house nearby — & soon after set up the trip-hammer [a large water-powered hammer]. But as Mr. H. was going one day to or from his office, he found that the apprentices had been round the pond in a boat knocking the frogs on the head and got a good sized tub nearly full of them. After that scarcely any were heard, and the trip-hammer being set up soon after, they all disappeared as if frightened away by the sound. . . . (Journal, July 31, 1855)*

Bullfrogs in Concord today may no longer have to fear marauding bands of apprentices trying to bash them in the head, but in the modern era frogs, toads, and other amphibians must face other more serious challenges to their survival. Despite our best conservation measures, the vernal pools that are so essential to the breeding of amphibians and other aquatic creatures could be threatened by climate change. Temperatures just a bit warmer could dry up these ponds in the early summer before the amphibians have completed their life cycle, causing the death of all juveniles for that year. This is particularly true for spotted salamanders, which have a long development time.

But before long-term climate change warms up and dries out vernal pools, they risk destruction by a whole range of other human activities and the pollution associated with them. Degraded and polluted pools are deadly for spotted salamander larvae. Imagine the harm to salamander larvae if the Concord golf course began to apply heavy doses of weed-killing chemicals and insecticides to kill the weeds and pest insects, and the chemicals run off the fields into the vernal pool; the juvenile salamanders would all die. Roads or other human structures that block access to the ponds can reduce salamander populations, as the adults would not be able to reach the pond to breed. And the surrounding forest itself needs to be protected since the adults need a place to live for the rest of the year. Protecting vernal pools won't help salamanders if we allow the surrounding forest to be cut down and developed for housing, golf courses, and other uses.

So we need to protect both the salamanders' wetland and forest habitats from immediate threats of human activity at the same time that we need to consider measures to offset the threat from long-term climate change. Only then can we hope to ensure a future for salamanders, species that have been around on Earth for over 150 million years.

Amphibians like these spotted salamanders might well be another class of species, like plants and birds, that have been changing the timing of their life cycle because of climate change, but we aren't going to find out anytime soon. The data from Massachusetts and maybe anywhere in New England simply aren't available. No one has put in the extensive time and effort needed to follow amphibians year after year, decade after decade, to see how their populations are faring and how the timing of their breeding habits have changed (or not) over time. Moreover, even if we had data showing population declines among salamanders, it would be extremely difficult to parse out whether climate change was to blame or the kind of habitat destruction described in the earlier paragraph. Amphibians are notoriously sensitive to even small amounts of pollutants, and a drop in population size could relate to multiple factors, either independently or in concert. We would need decades' worth of observations that track not only the animals, but the year-to-year changes in the local landscape, before we could discern if pollution, habitat loss, or temperature increases were the cause of any population declines. In other parts of the country, damaging parasite infestations, linked to pollution, are also harming many amphibian populations. And in the tropics, a widespread and expanding fungus is decimating frog populations. Amphibians are being hit from many directions, which explains why they are so endangered. Climate change may soon emerge as the latest and most damaging threat.

I'd love nothing better than to be able to analyze a mountain of observations regarding the breeding cycles of amphibians like the spotted salamanders of Concord so I could add it to what I've already discovered about the effects of Concord's earlier springs on plants and birds. The truth is, though, I don't yet have that kind of data, and probably never will. Well, that's the nature of science — discoveries often don't come easily or quickly, especially when the subject is so elusive. To obtain this type of data would require a long-term monitoring program supported by significant research dollars. And while Concord is notable for its amazing plant and bird data, we would have a hard time arguing that it is a hot spot for amphibian research.

Biology has come such a long way since Linnaeus first devised taxonomic classifications over two hundred years ago; now, instead of looking at a plant or animal and identifying it based on how it looks, we can examine the very molecules of its genes to assign it to one species or another. And yet there's still so much we don't know, and sometimes the old-fashioned observation methods of Thoreau, arduous and time-consuming though they might seem, are the best methods for finding out new data. Realistically, though, even if I were a trained herpetologist (which I'm not), I could probably spend the rest of my life studying salamanders in Concord and Newton and still not find the answers to my questions about the numbers, reproductive cycles, and migration times of the blue- and yellow-spotted salamanders. I can speculate about it all I want, but the salamanders won't be revealing their secrets to me anytime soon in the same way we were able to learn about plants and birds. So for now, reluctantly, I have to let go of these animals as a potential data source for climate change research. But I can still hope that some future biologist will take up the challenge and confirm (or refute) my speculations on the impact of climate change on Concord's spotted salamanders and other amphibians.

Hope for the threatened amphibians of Massachusetts and rare native reptiles like Blanding's turtles can only come if we introduce the next generation to the wonder of these secretive and shy creatures. As Regosin explained:

> Spotted salamanders captivate the imagination. I think that what makes these animals so striking is that they appear almost magically each year, seemingly coming out of nowhere on a few rainy nights in spring. The rest of the year these animals are underground in the forest, often confounding even the most experienced naturalist, who might spend an entire day turning over logs and still not find one. Although scientists may have learned much about this species, there is much we still do not know about these mysterious creatures.

It will take education and dedication to ensure that, when the next generation of field biologists goes out into the woods of Massachusetts, there will still be spotted salamanders for them to find.

Boston Marathon runners approaching Kenmore Square; photo by Noah Reid.

*He was a good swimmer, runner, skater, boatman, and would
probably outwalk most countrymen in a day's journey.*

RALPH WALDO EMERSON, DESCRIBING THOREAU, 1862

13. Running in the Sun and Rain

RUNNERS KNOW FIRSTHAND HOW WEATHER on race day affects athletic performance. Running a marathon of 26 miles, 385 yards places extreme demands on the body, demands often incompatible with hot weather. The Chicago Marathon of 2007 was stopped three and a half hours into the race after one runner died in the 88-degree temperatures and forty-nine runners needed emergency medical treatment. More recently, before the 2012 race, Boston Marathon officials advised inexperienced runners to postpone their participation until next year because of the predicted mid-80-degree temperatures. Of 26,655 registered runners in the 2012 race, 4,170 chose not to run and over 800 more did not finish the race. Even the man who won the Boston Marathon in 2011, Geoffrey Mutai, dropped out due to leg cramps brought on by the extreme heat.

Back in 1970, I was an innocent, idealistic Harvard sophomore living in Cambridge. While I was hardly a transcendentalist, I did believe that people and nations should live according to principles of moral behavior. And like many of my generation, I was having trouble understanding why the United States government was involved in the war in Vietnam. The Harvard campus was being disrupted by weekly antiwar demonstrations, strikes, even riots, and my evenings were increasingly devoted to political meetings. The chaos on campus was mirrored in my personal life as I struggled to understand both a confusing romantic relationship and my parents' financial problems, a situation that threatened bankruptcy and the loss of our family home.

In early February, I decided that I needed a simple personal goal to relieve my anxieties. I decided that I would run the Boston Marathon. Forty years ago the Boston Marathon was not the world-renowned athletic event it is today. When I decided to enter the race in 1970, the marathon still drew mainly regional athletes, attracting only a few elite runners from the international ranks. Despite being serious about soccer and track in high school, I had never run more than a few hundred yards in my life. In those days, distance running was not a normal part of athletic training; you trained for soccer by playing soccer, and you trained for sprinting by sprinting. So I began training for the marathon by running every day in the freezing, late winter weather, mostly around Fresh Pond in Cambridge, and each day I ran a bit a longer.

After eleven weeks of training and as the day of the race loomed ever closer, the longest distance that I had run was twelve miles, from my undergraduate dorm in Cambridge, along snowy streets, to our family home in Newton Center and back. Despite my obvious lack of preparation, I remained determined to run the race. Patriots' Day dawned overcast with a freezing drizzle. My father advised me to eat light (what did he know about running marathons?), and I ate a fried egg for breakfast, not the serious carbohydrate packing of today's runners.

After my father dropped me off in Hopkinton, I received my numbered bib and joined the huge milling crowd of runners, all men (officially, at least), and mostly in their twenties. When the starting gun sounded at noon, a deep roar went off in the crowd, signaling the release of the top runners in the front, but ten minutes went by before the runners around me at the back of the pack started to walk and then break into a run.

I felt good as I ran lightly along the course, but I was not prepared for the cold and the rain. I had dressed in shorts, a light shirt, and a windbreaker. I was cold. But my exhilaration kept me going. For the first two and a half hours or so of the gently undulating course, I moved easily.

Then I came up against Heartbreak Hill, so-called because it broke the hearts of the idealist young runners like me, confident in their ability do anything. Heartbreak Hill begins near the twenty-mile mark at Newton Center and continues for more than a mile past Boston College, and on to Cleveland Circle, a total ascent of eighty-eight vertical feet. As it had with generations of young men before me, the hill stopped me in my tracks. My muscles, fueled only by that measly fried egg, were starved of glycogen. I hit the "runner's wall" and could run no more.

I trudged up the hill in the freezing rain, my hands and arms numb with the cold. I was exhausted but not defeated; at least I was still moving forward. Now I could better see the dense crowds on either side of Commonwealth Avenue, spectators in heavy raincoats huddled under umbrellas. As I moved past the stone buildings of Boston College on my right, I heard a small boy ask his father, "Daddy, aren't they cold?" To which the father replied, "No, they are warm from running."

In contrast, the 2012 Boston Marathon took place against the backdrop of a coast-to-coast heat wave. Entrants in the marathon worried not about the cold but heat exhaustion. "Heat Forces Boston Marathon Dropouts," reported *ABC News*; "Heat Proves Tougher than Heartbreak Hill," said the *Washington Post*. In an online statement, the race organizers took the unusual step of warning runners to take it easy. "This will not be a day to run a personal best." Despite the warning, record numbers of runners had to seek medical attention for heat exhaustion and dehydration.

Six years earlier, Abe Miller-Rushing and I, along with a few other colleagues, had decided to look at the Boston Marathon as another classic sign of spring that might provide insights into climate change. We wanted to know whether rising temperatures were affecting race times. But could sports in Boston really tell us anything about climate change?

Athletic Endeavor and Climate Change

Abe and I knew that the effects of climate change were already being felt in the arena of winter sports. Opportunities in New England (and other northern states and countries) for outdoor ice-skating, snowmobiling, ice fishing, skiing, and other winter sports have become less predictable with bouts of unexpected warm winter temperatures. In some years, when abnormally heavy precipitation coincides with changing patterns of cold air movement, there is too much snow and ice, but in others there isn't enough — or we have a winter truncated by a warm autumn and an early spring. In that case, it has major economic consequences; ski resorts cannot attract skiers if their trails are not covered with snow. And the resorts can only make artificial snow if the air temperature is cold enough for the water to freeze and stay frozen. Throughout North America and Europe, ski resorts are struggling financially as they deal with shorter ski seasons and fewer skiers.

But are warm-weather sports also affected by climate change? In summer, high school and college teams begin practicing football, soccer, and cross-country for the coming fall season. With practices taking place under the blazing sun and in such high temperatures, it is not surprising that dehydration and heat prostration should be a frequent occurrence. In addition to students getting sick and collapsing, approximately four to five young football players across the country die each year from dehydration and the related effects of heat, according to the "Annual Survey of Football Injury Research." Such medical emergencies may become more common as our climate continues to warm. A warming climate is also of great concern in the demanding sport of marathon racing, where runners are in danger of overheating.

Few people know this as well as Bernd Heinrich, a professor at the University of Vermont and an extreme runner. On one race day in August 1983, the temperature was in the 80s. Heinrich was in Brunswick, Maine, at a track meet where he hoped to set an American record for one hundred miles. "I knew there was no chance [because of the heat], so I decided to go for the record for the greatest distance run in twenty-four hours. By going at a slower pace I would generate less heat, and I knew I would run half the distance at night, when it was cool." Heinrich's plan worked, and he set the American record for distance run in twenty-four hours: 156 miles, 1,388 yards, in temperatures that reached the 90s. At the race's end, Heinrich collapsed, and he spent the next twenty-four hours at a hospital in intensive care.

A biologist by profession, Heinrich has spent his long career studying thermoregulation in insects. Much of this work involves understanding how animals can survive during the long New England winters. But he also knows firsthand about thermoregulation in humans as a long-distance runner. He won the Masters Division of the Boston Marathon in 1980, just after he turned forty, but his passion is the ultramarathon. Technically, ultramarathons are any race longer than the 26-mile marathon. Heinrich has run, and won, races of 60 miles or more. He has even set the American records for distances of 100 km and 100 miles. As a result, he has unique insights, both professional and personal, into the demands that heat makes on the bodies of marathon runners.

Heinrich told me that in his experience, peak running times go down as temperatures rise and that hot days have specifically plagued the Boston Marathon. But he doubted that Abe and I would be able to find a statistically significant correlation, because of the high number of conflicting variables.

> One, training is now much more serious than formerly — so it could be that running times are actually getting faster as time (and temperature) increase. Two, maybe people are now more aware they need water along the way on hot days. Maybe with adequate hydration the temperature effect is masked.

Heinrich is sure of one thing: if dogs ran the Boston Marathon instead of humans, they would definitely show the effects of heat. "I can easily outrun our yellow Lab on a moderately hot day that feels very comfortable to me, but on a cold day he can run circles around me. On a hot day where I can run for at least several miles, he is panting before a quarter mile. Speed is very dependent on temperature." Heinrich's observation is explained by the fact that people have a special ability to lose body heat during hot weather by sweating from their entire skin surface; it is the evaporation of the water from the skin surface that provides the main cooling effect. In contrast, dogs are unable to shed as much body heat because of the insulating effect of their fur and soon become overheated.

In his 2002 book, *Why We Run: A Natural History*, Heinrich explains his theory of how evolution shaped the human body for long-distance running. The key feature is that with less body hair and the ability to sweat profusely, people were able to run down game during the heat of the day, without overheating. In contrast, other carnivores, such as big cats, like lions and tigers, and wild dogs, including hyenas and wolves, lose heat primarily by panting from their mouths. In the same warm conditions, these carnivores overheat and stop running.

But no matter how well suited our bodies may be for long-distance running, marathons take a toll: we deplete our energy stores and our muscles become damaged and overworked. But the biggest threat to the marathoner is heat, in the form of heat exhaustion or heat stroke and dehydration. Marathon runners face a constant challenge of getting rid of the enormous body heat generated by running. This heat is normally released through the skin and by breathing, and is enhanced by sweating.

However, when the temperatures are very warm, the body increases the rate of sweating in an effort to lose more heat. If this is not sufficient, particularly on hot days, body temperature can start to increase over normal levels, leading to a series of symptoms known as heat exhaustion or, when severe, as heat stroke, and include fatigue, cramps, heavy sweating, rapid breathing, fainting, and elevated body temperature. Needless to say, someone suffering from this condition is going to stop running and may need immediate medical attention. The loss of water from sweating can also lead to dehydration, characterized by thirst, nausea, headache, and unconsciousness. People suffering from heat stroke and dehydration need to be moved to a cool, shady area and be given cold drinks if possible. Under the extreme conditions of a marathon and without medical treatment, overheated runners even run the danger of dying.

Does a Warming Climate Affect the Boston Marathon?

Given how crucial temperature is to a marathoner's ability to finish a race, Abe and I were hopeful that by analyzing Boston Marathon times and temperature data spanning decades, we could gather more evidence for climate change in New England.

The Boston Marathon was first run in 1897, making it the oldest continuously run marathon in the world. Since 1924, the course has remained unchanged, and the date has remained more or less fixed as Patriots' Day. A Massachusetts state holiday commemorating the Revolutionary War battle of Lexington and Concord, Patriots' Day was originally observed every April 19 but later changed to the third Monday in April. The traditional start time of the race was fixed at noon, until it was shifted in 2007 to 10 a.m. to accommodate larger numbers of runners (by 2012 over 26,000) and a more diverse field of runners that included both men and women, seniors, and wheelchair racers. The earlier race time also made it possible for slower runners to finish in daylight hours and traffic flows to return to normal by evening.

Because of its long history, its fixed starting time, and stable route, the Boston Marathon is uniquely suited to examine whether a warming climate is affecting the race times of runners. So Abe and I came up with two research questions: Are the winning times of the marathon affected by the weather? And if so will a warming climate help or hurt the winning times?

Abe was able to obtain weather data for the two hours from noon until 2 p.m. on Marathon day from the nearby Blue Hill Meteorological Observatory for numerous weather variables, including temperature, wind direction, wind speed, humidity, cloud cover, and precipitation. Abe also used the wind speed and temperature numbers to calculate windchill, a commonly used index of how rapidly the weather removes heat from a body. We found that of all of the weather factors, temperature and wind are the only ones that affect winning times; the other factors need not be considered. However, we did have to include in our analysis the trend for faster races over the past century as the race became more international and better equipment and training methods became available. Over the past eighty years, men's winning times have decreased by about twenty minutes.

Let's consider temperature first. Most Boston Marathon records have been set on cold days when the temperature on race day is below 55 degrees Fahrenheit. Race times slow as temperatures rise. Finishing times tend to be slower by a full eleven seconds for each 1 degree rise in temperature. Considering that temperatures on the day of the marathon can vary from 32 to 90 degrees, this could change the time of the race by as much as ten full minutes, the difference between a record-breaking time and a relatively slow pace.

The effect of temperatures is clearly illustrated by the past two years of the Boston Marathon. The temperature in 2012 reached a blistering 89 degrees by the day's end, while on race day in 2011, the temperature at 1 p.m. was a more mild 57 degrees. The men's winning time of 2 hours, 12 minutes by Wesley Korir from Kenya in the extremely hot marathon of 2012 was nine minutes slower than the course record time of 2 hours, 3 minutes set during the cool running conditions of the 2011 race. Similarly, the 2012 winning time for women of 2 hours, 31 minutes by Sharon Cherop, also from Kenya, was nine minutes slower than the winning time of 2 hours, 22 minutes by Caroline Kilel in the 2011 marathon.

The Boston Marathon of 1976 had the hottest temperatures on record, with the thermometer hitting an unbelievable 90 degrees during the race. The winning woman of 1976 was Kim Merritt, who finished in

2 hours, 47 minutes, which is 2 minutes and 31 seconds faster than we would have predicted based on the weather conditions. Perhaps Merritt had a physiology that somehow allowed her to dissipate body heat more efficiently than other runners. However, a more likely explanation is that so few women were running marathon races in the 1970s that the potential times women runners were capable of was still being determined.

Abe found that wind also plays a factor in race times, though the effect he uncovered was smaller than that for temperature. It seems intuitively obvious that fighting a headwind during the Boston Marathon would slow down winning times. The data bore this out. Winning times slow by 21 seconds for every 1 meter per second (or 10 seconds for each mile per hour) increase in headwind speed. And athletes run faster — by the same amount — when they have tailwinds pushing them along. Because average wind speeds are often around 4 meters per second (10 miles per hour) on race days, race times could be affected by as much as 168 seconds, or almost three minutes, depending on whether there was a moderate headwind or a moderate tailwind. Considering both wind and temperature, we expect races run on cold days with a tailwind will be measurably faster than those run on hot days with a headwind.

Such cool conditions with a tailwind were exactly the conditions that occurred during the Boston Marathon of 2011. In fact, the winning time of 2 hours, 3 minutes, 2 seconds by Geoffrey Mutai of Kenya was the fastest recorded marathon ever run by any runner anywhere in the world. Despite this amazing feat, this time is not recognized as a world record. The International Association of Athletics Federations has ruled that the Boston Marathon route is ineligible for world records because it is not a loop and so does not account for a persistent tailwind. Also, the course trends slightly downhill, as the course starting point in Hopkinton is at a higher elevation (about 410 feet) than the ending point in Boston (about 19 feet).

Climate change could also be factor in Boston Marathon races. We know that Boston's average temperature has increased about 4 degrees Fahrenheit since 1893, due to both global warming and the urban heat island effect. And we know that long-distance athletes are not able to run as fast on warm days. Could warming temperatures on race days be causing the Boston Marathon to be slowing down? As it turns out, the temperatures on the specific day of the marathon have not gotten any warmer since 1893, because there is so much variation from day to day in Boston weather (particularly in spring) that it's hard to discover a warming trend over time when looking at a single day. We can only see the trend in warming when we use average temperatures for the year or the average temperature for a single season or month, not when we consider a single day.

This is consistent with Thoreau's diaries, which also reveal the extreme variation in daily weather in New England, as we can see on two contrasting Patriots' Day observances. On April 19, 1855, Thoreau wrote, "Warm and still and somewhat cloudy. Am without greatcoat. The guns are firing and the bells ringing [for Patriots' Day]. . . . Many tortoises have their heads out. . . . The wild red cherry will begin to leaf tomorrow." Contrast this with his journal entry from the same day in 1854, the previous year: "This is the fifth day that the ground has been covered with snow. There first fell about four inches on the morning of the 15th. . . . Then as much more fell on the 17th, with to-night the ground is still more than half covered. There has been sleighing. . . . Farmer says that he saw a man catch a bluebird yesterday which was dying in the snow." We can see that even though there was great variation in the temperature on Patriots' Day in Thoreau's time, he saw variations from wintry to pleasant, where today we tend to see variations from cool to hot.

Computer models now tell us how much the temperatures in Boston are expected to increase by the years 2050 and 2100. Abe has used these projected temperatures to estimate future winning times for the Boston Marathon. His simulations suggest that by 2050, warming temperatures will have about a 30 percent chance of affecting race times, but by the year 2100 there is more than an 85 percent chance that warming temperatures will slow down winning times. However, that is only if race times continue to be run at midday. In fact, starting in 2007, the start of the race time was shifted to 10 a.m. But it is likely that even marathons with earlier start times will slow down as temperatures warm in coming decades. And that is what happened in 2012. Even with the 10 a.m. start, both the men and women's winning times were slower than normal due to the exceptionally hot conditions.

In future years, climate change will likely have numerous effects on sports activities in New England, elsewhere in the United States, and around the world. Ski seasons will get shorter, and skiers and ski tournaments may have to travel farther north for reliable conditions. Ice fishing will also have a shorter season and might become more dangerous due to thinner ice. Athletes in warm-season sports such as soccer and football may need to avoid outdoor practices during the middle of summer days to avoid dehydration. Given the 35,000 deaths linked to heat waves in Europe in the summer of 2003 and the heat-fueled wildfires in Russia in 2010 and the American West in 2012, heat-related changes to sports may seem insignificant by comparison. But such changes to sport schedules, in which many people participate and many more are spectators, offer an opportunity to educate millions of people about climate change.

Ultimately, the Boston Marathon and other events like it provide us with a highly visible snapshot of how temperature is going to change our ability to work and play in a warmer world. Warming temperature will have severe impacts on farmers, construction workers, forestry workers, and others who have to work outside. For them, a warming climate will have significant costs if they have to shorten their work hours on particularly hot days or if they work anyway and become exhausted or sick.

The Other Side of Heartbreak Hill

And how did I finish my own Boston Marathon on that cold, rainy day in April 1970?

Upon reaching the top of Heartbreak Hill, I somehow found the strength to once more break into a run. And with a slow jog interrupted by walking, I conquered the next low hill leading to Cleveland Circle. And from there, I ran slowly all of the way to the end of the course at the Prudential Center in Boston, finishing in just under four hours.

My friend Marty, whom I had known my whole life, met me at the finish. When I entered the rest area for runners, Marty did not recognize me. I was so exhausted, weary, frozen, and otherwise altered by the experience, that in his eyes I was a different person. And I *was* a different person. In the days and weeks that followed the race, I left the confusion and uncertainly of the past months behind. I still did not know where I was going in my life — that would only become apparent the following year when I discovered botany — but I was ready for whatever was next.

In coming decades, climate change will challenge us all physiologically and in many other ways, whether we are marathoners or not. All of us will need to change our game plan to meet those challenges. The question then becomes, how will we cope with climate change, locally and globally? Will we keep doing what we've been doing and hope we can survive in an increasingly warm world just on our sheer will to keep going? Or will we begin now to plan and strategize ways to conserve our energy resources, and slow or even stop the warming trend? Ignoring climate change is a tempting option, especially when there are political leaders saying that other funding priorities are more important, such as creating jobs, improving schools and health care, buying new weapon systems, or fighting terrorism. Such a call to action against terrorism takes on special meaning for the people of Boston because of the bombing attack during the 2013 marathon. But even though the country is probably capable of coping with the short-term consequences of climate change, the longer-term effects of warming temperatures, rising sea levels, and changing precipitation patterns could devastate the country by the end of this century.

In a way, our current situation in regard to climate change is a lot like a marathon. We're at Heartbreak Hill, slowed down by the enormity of the task ahead and many competing priorities — but we have to keep going, or the race (for us) will be over. If we persevere, however, and make it through to the finish, who knows what we'll be capable of?

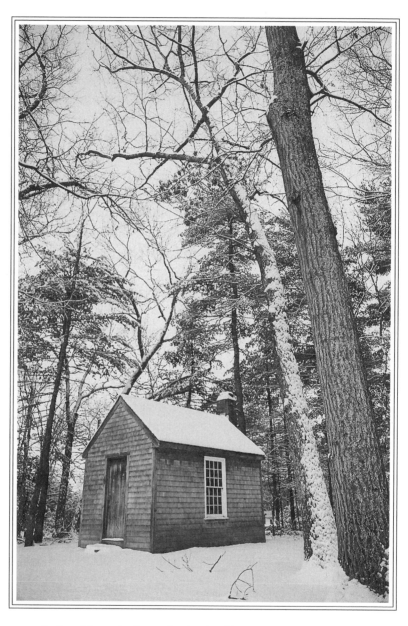

Replica of Thoreau's cabin on the edge of Walden Pond; photo by Tim Laman.

We cannot see anything until we are possessed with the idea of it,
take it into our heads — and then we can hardly see anything else.

THOREAU, *AUTUMNAL TINTS*, 1862

14. A New Earth

SUPPOSE, HALF A CENTURY INTO THE FUTURE, in the year 2064, Henry David Thoreau returned to the edge of Walden Pond. What would he find? There are a number of possible scenarios. Let us follow him on an imaginary walk through Concord and Boston, seeing through his eyes one of these possible futures in which the world continues on its present warming trajectory.

Thoreau would notice many changes. The winter weather is no longer as cold as it used to be; the daytime temperatures are mostly in the pleasant 50s, and the nighttime temperatures are above freezing. Walden Pond no longer freezes over, and wood ducks remain on the pond all year long. Looking at his almanac, he realizes that Concord now has a climate like South Carolina had back in his own time. On a trip to the White Mountains, he observes many abandoned settlements; local people tell him that these were formerly ski and winter sport towns whose economies collapsed when the weather became too warm for a long snow season. The weather is also too warm for tapping sugar maple trees.

In the spring, there are hardly any wildflowers in the forests; the climate is now suited to more southern species, but they have still not migrated to the area. The trees are leafing out in late March rather in than in early May, and many seem sickly because of attack by newly arrived invasive insects. And there are not as many migratory birds as he remembered. Walden Pond has become too warm for trout. The swamps and bogs where the naturalist used to enjoy spring choruses of peepers and other frogs have now dried up and are silent.

On a trip to Boston, Thoreau observes a new six-foot-high flood wall, all along the border of the Boston Harbor facing the sea, built to keep out the high tides and storm surges associated with more frequent nor'easters and hurricanes and a rising sea level. In the harbor itself, lying on either side of the main channel, Thoreau is dumbfounded to see two massive steel structures, taller than the Washington Monument and each more than a quarter mile long. These are floodgates designed to close and sink during storms, protecting Boston from flooding, including the low-lying lands of the Charles River basin and Mystic River basin. The barriers were built at a cost of over $5 billion. Many other towns and coastal areas in Massachusetts have been overwhelmed and destroyed by flooding and their populations moved inland because it was too expensive to spend billions of dollars to protect them. Cod and lobster are no longer found off the Massachusetts coast; the water here is now too warm and acidic, and they are only found in small numbers in northern Maine.

During four trips to Cape Cod from 1849 to 1857, Thoreau observed the powerful influence of the sea on the landscape and people, which he recorded in his book *Cape Cod*. Here in the middle of the twenty-first century, Thoreau learns that many of the homes of Cape Cod have been damaged and destroyed in a series of powerful hurricanes, and the outer cape has been broken into islands by erosion and storm surges that overwhelmed the barrier beaches. Lastly, the damage to the water supply caused by decades of overdevelopment combined with a rising sea level forced people and businesses to leave.

In Concord and throughout Massachusetts, the summer heat is oppressive, with day after day above 100 degrees, and nighttime temperatures lingering in the mid- to high 80s. People avoid going outside during the middle of the day. The coming of autumn is no longer heralded by cool, crisp mornings; the weather is warm, and the leaves change from green to brown and simply drop from the trees, without creating the tapestry of brilliant color for which New England was long famous.

An Alternative Future

This is just one possible scenario for fifty years in the future, one that assumes we continue on our present course of rapid demographic and economic growth. If the already-excessive human demands on nature continue to grow, the world continues to consume coal, oil, and gas at an increasing rate, and tropical countries continue to cut down rain forests, carbon dioxide levels in the atmosphere will continue to rise at a rate of 2 parts per million per year and reach close to 500 parts per million by 2064. As a result, the climate will likely warm by as much as 6 degrees Fahrenheit, and sea levels will rise by two feet or more due to the melting of polar ice and glaciers and to thermal expansion, the property of seawater that causes it to expand as it warms.

But we can imagine a very different scenario in which the problems of global warming and climate change have been recognized as an international emergency threatening the world's water and food supplies, tropical and temperate forests, biodiversity, and the very existence of coastal cities. The nations of the world in this future have recognized limits to growth, and responded by implementing a wide range of measures that substantially reduce the consumption of fossil fuels and the consequent production of greenhouse gases, and restore tropical and temperate forests around the world.

Let us now walk beside Thoreau through this very different future landscape. Thoreau again returns to Concord on an early January day fifty years from now, but even though the weather is somewhat warmer than two centuries ago, Walden Pond still has a layer of ice, and the philosopher can't stay in his unheated cabin because the temperature dips below freezing every night. Almost every house in Concord has solar panels on the roof to generate electricity and to heat water. New houses average about half the size of present houses and are built with advanced insulation and thermal windows to minimize heat loss. The people in Concord mostly travel by bicycle and public mini-bus; the number of private cars is far fewer than today. Both smaller and lighter than today's typical cars, the cars of this future brim with energy-saving technology.

During a visit to Boston, Thoreau observes extensive new train systems and other forms of public transportation. The hills and harbor areas are dotted with wind turbines for generating electricity, and new buildings are covered with a photovoltaic surface that feeds electricity into the power grid. There are new flood walls along the edge of the harbor, but they are only three feet tall, and the government decided that building the giant flood gates was not necessary. At a public meeting, plant scientists describe vast projects to replant tropical and temperate forests, both to absorb atmospheric carbon dioxide and to produce wood and paper products. These scientists are also developing new varieties of plants to produce food more efficiently — especially plant protein, now that almost everyone is either a vegetarian or has cut back on meat consumption for health and environmental reasons. The current buzzword is "enhanced moderation," sometimes shortened to "enmo," to indicate the goal of individuals and society to enjoy a rich cultural, intellectual, social and public life without using a lot of resources or generating additional greenhouse gases. There is even a "Henry Society" of people who live as simply as possible and spend their time learning about and caring for nature.

These are only two possible scenarios that Thoreau might face if he returned to Walden Pond half a century from now. It is up to the citizens and leaders of today to decide which path we will follow. People can start to take individual actions to reduce the production of carbon dioxide, but they also need to pressure their governments to mount large-scale efforts against climate change through political action: lobbying, voting, and demonstrating. If we take no action, the first scenario, with all of its unintended consequences, will happen on its own.

Is Climate Change Really Happening in Concord?

At the start of my research project in 2003, a skeptic could justifiably claim that there was little or no evidence that a warming climate was affecting the plants and animals of Concord and the surrounding area of Massachusetts in terms of the abundance, distribution, and timing of spring activity. The same lack of solid information held for the neighboring states of New England and indeed most of the entire eastern United States. Over the past twelve years, however, dozens of my colleagues, collaborators, and students have provided abundant evidence that the warming of eastern Massachusetts, due to both global warming and urbanization, is linked to the earlier thawing of Walden Pond in late winter, the earlier flowering times of local plants, the changed flight times of butterflies, and even the different arrival times of some bird species. And in a truly remarkable finding, we learned that a warming climate is responsible for the decline and local extinction of well-known wildflowers. The last three years have provided even stronger evidence of the effects of climate change. The spring of 2010 was the warmest in the past 180 years in Boston, and 2012 had the warmest first three months of the year and among the warmest springs. As a result, plants flowered three weeks earlier than in Thoreau's time. In the remarkably warm winter of 2012, Walden Pond had only formed a thin skin of ice by January 19. Ice-out occurred two weeks later on January 29, while in Thoreau's time ice-out varied from mid-March to mid-April.

These changes in the climate and environment of Concord are a local version of what is happening to the whole planet. Global temperatures have already increased by an average of 1.3 degrees Fahrenheit over the past century, with the four warmest years being 1998, 2003, 2005, and 2010. For the lower forty-eight states of the United States, 2012 was the warmest year ever recorded; according to the National Climatic Data Center, 38 percent of the lower forty-eight states were suffering from severe to extreme drought in 2012 due to a combination of record-breaking hot temperatures and low rainfall, causing major losses to livestock and crops. The U.S. Environmental Protection Agency website reports that sea levels have risen by eight inches over the past 140 years, and that this rise has been accelerating in recent decades. Even the vast ocean has become 30 percent more acidic over the past 250 years, as more carbon dioxide dissolves in the water. This has potentially devastating consequences for marine animals that need precise sea chemistry to live and to grow their skeletons.

But some say "so what?" What is so wrong about this warming climate? As a native of New England, I have experienced the winters of our region for the past six decades, and I know that most New Englanders would probably welcome the idea of milder winters with fewer subzero mornings and correspondingly longer summers. One problem with a warmer climate, however, is that the plants and animals of Massachusetts are not suited to it. When conditions get a few degrees warmer, many of our wetlands will dry out in the summer, leading to the loss of endangered wildflowers and animals. Stressed by drought and the arrival of new insect and fungal pests from the southern United States and other countries, many of the trees in our iconic northern forests will die. This human-caused degradation of wild nature is deeply wrong.

In addition to harming nature, we will harm ourselves. The health of people, especially the very young and very old, will be damaged by day after day of exhausting summer temperatures over 100 degrees. But the most serious problem of all for residents of eastern Massachusetts will be rising sea levels, especially when driven by storm surges from more intense hurricanes and nor'easters that will either overwhelm cities and towns, or require billions of dollars to be spent on new seawalls and dikes.

The problems of climate change will be felt even more severely in other areas of the country and on other continents where agriculture is already difficult and water is scarce. Huge areas of the midwestern and southwestern United States will experience disastrous water shortages and droughts, leading to agricultural collapse and water rationing for cities and industries. The exceptional drought of 2012 will become a typical annual event. Many farming areas of South Asia, South America, Africa, and southern Europe will turn to scrub or desert or otherwise become non-arable, leading to starvation or the large-scale emigration for tens of millions of people. Flooding caused by rising sea levels will overwhelm coastal cities in many poor regions of the world. Climate change also threatens many rare and endangered plants and animals beyond New England. Many tropical species in particular are found in only limited areas; when the climate becomes hotter and drier in such places, these species may not be able to live there any longer and will become extinct. It has been estimated that climate change will threaten hundreds of species of birds with extinction by the end of the century. So, while the warming temperatures and earlier flowering times evident in Concord may be personally pleasant and even scientifically intriguing, they are really the first warning signs of an impending disaster about to hit the global environment and human society.

The World Is Changing for the Worse

In an intriguing passage in *Walden*, Thoreau describes a conversation with a woodsman, someone whom he believes to be more in touch with nature and simple truths than the average citizen of Concord. This woodsman works all day in the forest cutting trees and enjoys the simple pleasures that his surroundings provide him. At one point Thoreau notes:

I heard that a distinguished wise man and reformer asked [the woodsman] if he did not want the world to be changed; but he answered with a chuckle of surprise in his Canadian accent, not knowing that the question had ever been entertained before, 'No, I like it well enough.' It would have suggested many things to a philosopher to have dealings with him. (Walden, 160)

Even if we identify with the woodsman and want the world to remain unchanged for future generations and ourselves, this will not be possible. The world is changing in measurable ways. At the most basic level, the concentration of carbon dioxide in the atmosphere is increasing each year by about 2 parts per million, or about half a percent. This is a real change that can be quantified — and it is caused by our continuing use of fossil fuels and by the destruction of forests. As long as the concentration of atmospheric carbon dioxide and other greenhouse gases continues to increase, the world will continue to warm.

In his 2010 book of the same title, the environmental writer Bill McKibben coins the novel term "Eaarth" to call attention to the fact that we are living on a new, degraded planet that is likely to be harsher and uglier than the lovely "Earth" we have known. He also points out that politicians always talk about the need to deal with climate change for the sake of our grandchildren, because they believe climate change impacts might not be present until many decades in the future. Actually, we need to reduce the effects of climate change for our own children and ourselves because the effects of a changing climate are already present. If we cannot set limits to our own numbers and demands on nature, there may be no limit to how dysfunctional the "Eaarth" could become.

What Do We Need to Do?

What do we need to do to reduce our production of greenhouse gases and reverse the harmful effects of climate change? According to the Intergovernmental Panel on Climate Change (IPCC), the main drivers of climate change are population growth and rising per-person consumption of natural resources. In order to keep climate change from reaching disastrous proportions, humanity will have to shift from a pattern of always creating more wealth for more people, to an alternative path of providing sufficient resources for a limited number of people.

At present, global human activities emit about 8 billion tons of carbon gas each year, and this amount is projected to double in the next fifty years under "business as usual." To maintain our present atmospheric level of carbon dioxide and not slide into disastrous warming, the Carbon Mitigation Initiative at Princeton University, as described on their website, has identified classes of approaches that could be implemented to reduce carbon emissions. Each of these approaches is termed a "wedge," for its ability to reduce carbon emissions in small amounts right away and to "scale up" over time so as to reduce carbon emissions by 1 billion tons per year in fifty years. By implementing seven or eight such wedges, carbon emission levels could be stabilized at levels approaching those of the year 2000. The most promising plans are as follows:

Wedge 1. Doubling the efficiency of cars and trucks, in terms of miles per gallon of gas, by encouraging the use of smaller, lighter vehicles and improved technology.

Wedge 2. Halving the number of car miles driven, in part through improvements in public transportation.

Wedge 3. Using the best technology to more efficiently light, heat, and cool our homes and other private and public buildings.

Wedge 4. Increasing wind power production by a factor of 10.

Wedge 5. Increasing solar power production by a factor of 100.

Wedge 6. Replacing coal-powered plants with ones powered by natural gas or nuclear power.

Wedge 7. Protecting all remaining tropical forests, and replanting forests when possible.

Wedge 8. Capturing and storing the carbon dioxide produced by coal-fired plants.

The driving force for achieving the goal of these wedges could be a tax on greenhouse gas emissions that makes it cheaper to reduce emissions than to continue with current activities. Another approach is a fixed limit on greenhouse gas emissions allotted to each country, forcing nations to regulate their internal activities to be more efficient and to restrain their use of fossil fuels.

The weakness of the Princeton approach is that it does not directly address the underlying issues of increasing resource consumption and population growth. Without addressing these issues, this solution, by itself, seems unlikely to succeed.

Phil Cafaro, an environmental philosophy professor at Colorado State University, has proposed additional behavioral and institutional wedges that are feasible with present technology, that could reduce carbon dioxide emissions as much as the Princeton wedges, and that directly limit the growth in human numbers and personal consumption that are driving climate change. These include the following:

> *Alternate wedge 1.* Prevent half the growth in meat consumption and production projected to occur over the next fifty years. Meat consumption and its associated agriculture presently account for around 14 percent of world greenhouse gas emissions. This could be reduced through taxes on meat, ending agricultural subsidies, banning industrial feedlots, and encouraging more healthy plant-based diets.
>
> *Alternative wedge 2.* Hold civilian air travel and air freight shipping, one of the most rapidly growing sources of carbon dioxide, at current levels, through higher taxes or direct limits on the number of flights allowed per person.
>
> *Alternative wedge 3.* Reduce carbon dioxide production by individuals through luxury taxes on gas-guzzling cars, excessively large houses and apartments, and other unnecessary, high-emission consumption.
>
> *Alternative wedge 4.* End population growth by providing access to inexpensive birth control and family planning services, and improving the educational and economic opportunities available to women.
>
> *Alternative wedge 5.* Use monetary and fiscal policy to slow the economic growth that is the primary cause of increasing carbon dioxide emissions, rather than following the status quo, where nations work to increase growth as much as possible.

Perhaps some combination of efficiency improvements and institutionalizing the acceptance of limits would be most successful. All of these various wedges will be more effective if they are implemented sooner and strictly enforced. And given their vast scales, these actions will need to be implemented, regulated, and subsidized at national and international levels.

Various international meetings of scientists, political leaders, and environmental groups have been convened to develop the treaties and agreements to reduce greenhouse gas emissions, such as the United Nations Conference on Sustainable Development, known as Rio+20, held in Rio de Janeiro in June 2012. However, agreement on serious reductions has proven impossible to achieve, primarily because the world's leaders do not want restrictions on economic growth. Even modest national and international reform efforts to reorient economic activities to reduce the production of greenhouse gases are frequently defeated by well-funded corporate lobbying efforts, established to slow or block needed change.

Can Thoreau Show Us a Better Path?

In short, I am convinced, both by faith and experience,
that to maintain one's self on this earth is not a hardship but a
pastime, if we will live simply and wisely. (Walden, 75)

While we need to continue urging and demanding that our political leaders, governments, and international bodies take action, as individuals and as local communities we can start to address the problem of climate change in our own ways. Thoreau's advice in the 1850s to his fellow citizens of Concord would likely be the same if he were addressing a public forum today on the topic of climate change: that they and we should lead a more materially simple life, and in the process minimize our contribution to rising levels of carbon dioxide. From Thoreau's perspective, the key first step is a change of mind-set — one that takes us from regarding "stuff" as something we want more of, to one that looks upon unneeded belongings (that is, not necessary for keeping us alive and healthy) as burdensome. He observes, "When I have met an immigrant tottering under a bundle which contained his all. . . . I have pitied him, not because that was his all, but because he had all that to carry. If I have got to drag my trap, I will take care that it be a light one . . ." (*Walden*, 71).

There are actions that can benefit the world and can also help us on a personal level. We know that a simple diet of unprocessed foods (that does not require much fossil fuel to produce), exercise (rather than spending hours riding in a car or sitting in front of a computer or television), and a life with less stress and complications (from too many possessions) can also help secure a healthy and long life. According to the Mayo Clinic, these simple lifestyle changes mentioned above can lead to reduced chance of heart attacks, improved mental health, and better ability to sleep.

As Phil Cafaro eloquently states:

> In itself, one individual consuming less is trivial, in the context of global climate change. But that person freed from the desire for ever "more" is now in a position to ask for a new kind of politics from his leaders and his fellow citizens: a politics of "enough" rather than the current "more more more."

Living Simply in Japan

In the autumn of 2006, I had the opportunity to learn firsthand about the value of living simply and minimizing my impact on the environment during a three-month visit to Tokyo. Housing in Tokyo is famously expensive, and my host at the University of Tokyo had expended considerable effort to find me something suitable and economical. What he offered me was a simple one-room apartment in a dormitory for foreign graduate students. He was very hesitant to offer this apartment to me, because it was so small for a visiting professor, and even tiny by Japanese standards. I admit I was dismayed when I first set eyes on the apartment, a room eight feet wide by fourteen feet long. At the far end of the room there were glass sliding doors opening onto a small balcony that overlooked a city park. Next to these sliding doors was a single bed that could fold up against the wall. On the opposite wall was a simple desk and chair. As I entered the room, on my immediate left there was a tiny bathroom that measured about five by four feet, furnished with a miniature sink and toilet, and a shower that was just possible to enter and turn around in. The water for the shower came from a rubber tube attached to the sink faucet. Past the bathroom on the left was a kitchen area consisting of an electric hotplate, two square feet of countertop, and a small sink. A miniature refrigerator big enough for one or two days of food sat on the floor. Thoreau would have felt right at home.

Used to the comfortable existence of my middle-class home in the suburbs of the United States, I felt at first that this minuscule living space was intolerably cramped and claustrophobic. But I soon began to appreciate what a wonderful choice this had been. First of all, because this apartment was so small, the rent was very low and affordable, only $300 per month, whereas other visiting professors were renting regular apartments for five times my price at the going rate of $1,500–2,000 per month. Second, from an environmental point of view, I was using far less energy to heat and cool this tiny apartment than if I had rented a much larger residence. Third, I could give this small apartment a thorough cleaning in just ten minutes when I was expecting visitors, rather than the hour or two it took our family to prepare for guests back in the United States. Finally, by having such a small space, I was forced to edit my possessions down to only what I needed for my daily life; there was no superfluous clutter to get in the way. I had arrived with just one suitcase, consisting of a week's worth of clothes, winter boots and jacket, a few books, camera, computer, and some simple kitchen utensils, and that was it.

I enjoyed the simplicity of my new housing, and I soon identified with Thoreau and his cabin in the woods. Even though he lived in an isolated cabin and I was in an urban apartment, our living situations shared the experience of simplicity. I could relate to Thoreau when he remarked, "My dwelling was small . . . ; but it seemed larger for being a single apartment and remote from neighbors. All the attractions of a house were concentrated in one room; it was a kitchen, chamber, parlor, and keeping-room" (*Walden*, 264).

I was concerned that I would not be able to have my many Japanese friends over for dinner. Americans, and I'm no exception, generally like to entertain friends in their houses for evening meals. This is not true in many other countries, such as Japan, where most meetings among friends take place in restaurants or other public spaces. How could I invite friends to my tiny apartment for dinner? Again, the biggest obstacle was my own mind-set. My two chairs and the bed provided seating for four people; if I needed more chairs for a larger party, I just borrowed them from my student friends in the dormitory.

As for food at these gatherings, we just ate simple meals of rice and steamed vegetables and fish, cooked on my single burner. Often my guests brought additional things to eat, and this made for more of a shared experience. My guests were coming for the conversation and fellowship, not for any gourmet cooking. It was readily apparent that for my Japanese visitors, these were unique experiences, in terms of visiting someone's apartment for dinner, more particularly an American's apartment, and even more so because the apartment was so tiny. The equivalent in strangeness for Americans might be a winter party at someone's house in which guests are invited by the host to come in bathing suits. And yet these dinner parties in the tiny apartment worked, and we all had a great time.

My experience in Japan showed me the value of a more simple existence. I recognize that many people in the United States and in less-developed countries around the world are forced to live a simple life, dwelling in a small apartment or house and doing without a car out of necessity. At the same time, as we've seen since the economic recession that began in 2007, the hyper-consumer lifestyle of "bigger and better" can become a trap for many Americans. In the lead-up to the housing collapse, millions of people wound up buying bigger houses and taking on larger mortgages than they could afford, requiring both members of a couple to take on demanding jobs (or multiple jobs) just to make ends meet. Then when companies and the federal and state governments began laying off workers, many of these people lost their jobs, fell behind on their payments, and lost their homes. The suffering caused by all this was immense. As the housing market begins to recover, many potential home buyers are looking at smaller houses and smaller mortgages to avoid this trap.

Thoreau's Advice to Modern Americans

More than 150 years ago, Thoreau provided insights into the dangers of a society that is overly concerned with acquiring possessions, speaking of a "seemingly wealthy, but terribly impoverished class . . . who have accumulated dross but know not how to use it, or get rid of it, and thus have forged their own golden or silver fetters" (*Walden*, 16). Americans today are immensely wealthier than Thoreau's neighbors and fellow townsmen, yet we have hardly advanced at all in the wisdom necessary to use those riches wisely. Perhaps the time has come to work less on increasing our "means" and instead focus more on clarifying and elevating the "ends" of life.

Following Thoreau's advice to live simply and so avoid the pitfalls of hyper-consumption and "affluenza" — the unsatisfying and unending pursuit of material wealth — can help us deal with the problem of climate change. If we can be satisfied with a modest place to live and a modest lifestyle, rather than always striving for something newer, larger, and more fashionable, we will be in a better position to limit our demands on nature. This is one key to creating societies that have a chance to be ecologically sustainable.

Thoreau also urged his fellow citizens to be aware of the implications and consequences of what we do. Following his lead, we need to be mindful consumers. We need to be aware that when we drive large cars or take plane flights on vacations, we are using fossil fuels that affect atmospheric carbon dioxide levels. Our choice of what type of meat, fish, or vegetable protein we eat, what types of transportation we choose to use, and what types of homes we live in can impact not only our own health and personal budget, but can have implications for the health of the tropical rain forests, distant seas, and, ultimately, for climate change. We need to learn about these connections and adjust our habits accordingly.

Thoreau would also be an advocate for practical commonsense approaches to climate change, rather than elaborate technological fixes for the distant future. For example, the automotive industry is pursuing fuel economy solutions, predicting cars with high mileage will arrive sometime in a decade or two. However, more direct incentives for people to drive less or use less fuel could be implemented immediately if there was sufficient political will. Such incentives would include subsidies for small fuel-efficient cars, regulations that strongly favor carpools, discounts on monthly passes for public transportation, infrastructure that encourages the use of bicycles, and taxes for driving into city centers.

Thoreau also provides insights into how we can save energy in our homes, which presently accounts for 40 percent of U.S. energy consumption. His philosophy was to live modestly and also to repair and reuse his possessions rather than to buy new things. In our neighborhood over the past few years, many of the families are replacing their old windows at great expense in order to "save money" on heating and cooling. However, they could achieve the same level of savings at a tiny fraction of the cost by keeping their old windows in good repair and turning down the thermostat by a few degrees in the winter and up by a few degrees in the summer. Other neighbors are ripping up and replacing their old cracked asphalt driveways, to create new ones with a smooth, glossy black surface. But if they just fixed the cracks with a bucket of sealant, the old driveway would work just as well as the new one, and tons of debris would not be created. By following Thoreau's ideas, these neighbors would save tens of thousands of dollars and lower their impact on the environment.

One thing that made Thoreau so effective as a thinker and writer was his ability to gain insights from observation in the natural world. If our goal is to protect the environment and deal with the problem of climate change, then part of our strategy should be for each of us to immerse him- or herself in nature in order to understand what we are trying to protect. At the most basic level, this means walking through natural landscapes and observing what is there; we should learn the names and characteristics of the birds, mammals, plants, and other species. We should observe their behavior, their migrations, and their seasonal changes, to better understand their and our place in nature. As we develop this understanding, we will become better advocates for their protection.

THOREAU SPENT TWO YEARS AND TWO MONTHS living in the woods by Walden Pond and probably walked through a part of them for most of the days of his adult life, devoting himself to the study of nature and learning the lessons of living simply. He was also intensely engaged in the major issues of this time, including the abolition of slavery and opposition to the Mexican-American War. He wrote about these political topics, gave speeches, and even refused to pay his taxes to support what he felt were unjust government actions. Today we can follow Thoreau's lead in dealing with the problem of climate change. The abolition of slavery was the most important issue facing society in Thoreau's time, and climate change is arguably the most important issue that we must deal with today. At an individual level, we can choose to lead the most responsible lifestyle possible; but we must also work together within the political process to address the problems that we face as a society. Climate change is such a complex and far-reaching problem that it will require multiple collective actions by cities, states, and nations. Ultimately, all nations must work together to find the answer. The citizens of the world and their leaders will be more likely to take action when they can see the immediate effects of a changing climate. People might be convinced by an unusual heat wave or a summer drought such as the United States experienced in 2012, the rising sea that floods the coastal area where they live, or the melting of sea ice in a place that always had ice before. Or people might see the effects of climate change through careful study of the plants and animals in the surrounding landscape.

I became convinced of climate change when I saw blueberry bushes flowering in early April, a month earlier than Thoreau had seen them on the shores of Walden Pond a century and half ago; other evidence may be extremely varied, but it is present nearly everywhere. Climate change is happening now, and it is time to take action.

Afterword: Citizen Science

CLIMATE CHANGE IS ALREADY HERE. We can see its effects in weather patterns and from the dates when ice melts on ponds, trees leaf out, flowers open, and migrating birds arrive in spring. The timing of other natural phenomena will alter in coming years as temperature and rainfall patterns change and the sea levels rise. Climate change will affect every place in the world, though its effects will be felt more strongly in some locations than in others. And the particular effects of climate change will vary across the surface of the Earth.

Governments around the world have finally begun to mobilize to deal with climate change. Efforts have begun both to slow the warming of the planet and to adapt to the inevitable challenges of a changing climate. It is important that these efforts are based on solid data. It's here that ordinary people have a role to play as citizen scientists.

Before the professionalization of science in the nineteenth century, all scientists were citizen scientists, and ordinary folks often undertook careful study of nature. The long tradition of what we now call citizen science has given us such works such as Gilbert White's *The Natural History of Selborne* (1789). For forty years White, an English parish priest, kept a record of the natural phenomena around his home in Hampshire. His diary remains one of the most reprinted books in English, after the Bible and the works of William Shakespeare.

Following in the footsteps of Thoreau and White, you can keep your own climate change records. The best way to keep track of the effects of a changing climate is to make careful and regular observations of the environment where you live. This dedication is the key to Thoreau's success as an observer. He walked around Concord every day, typically for four or more hours, and afterward wrote up his observations while they were fresh in his mind and he could still recall details. You don't need to spend four hours every day making observations; the key is to make observations, even briefly, every day or at least several times a week.

In the coming decades, people across the United States and throughout the world will see the effects of a changing climate in their backyards and gardens. You will notice these effects as you go on your favorite walks through nature reserves and public gardens, as you commute into work, and as you walk, bike, or drive through the country or just in your local neighborhood. To be useful scientifically, your observations should be recorded in a journal. Such a journal can be kept throughout the year or just during the spring or another season when the pace of change is most rapid and most clearly driven by climate. Perhaps there is a lilac bush or a weeping willow in your yard, in a favorite park, or next to your workplace. When does the lilac open its first flower, and when is it in peak flower? When does the weeping willow produce its first flush of leaves? And once you make such observations for a few years, you will start to see patterns. Does the lilac flower the same time each year? Does the weeping willow have its first flush of leaves the same day each year? Is the timing affected at all by the weather that year?

Getting Started

Once you decide to keep a nature journal, the first step is to learn to identify the trees, shrubs, and wildflowers, the birds and insects, and other creatures that live in your area. Inexpensive field guides, available from bookstores and public libraries, are an excellent resource for field identification. Websites such as allaboutbirds.org offer help with identification, and with the advent of smart phones, dozens of apps have been developed to help with identifying wildlife. Regular observations are important to success: record what you see every day if you can or once or twice a week. Like Thoreau, record your observations in a journal soon after the end of the session, while the details are fresh. A digital camera or sketchbook can help. Your first year of observations might have some errors until you become confident in your identifications, or if you do not know where to find a particular plant, insect, or bird. But with practice, you will soon be making valuable observations that can help document the effects of climate change.

You can tailor your nature journal to any location: spring flowers in the eastern United States, the arrival of different species of hummingbirds in the Southwest, or the appearance of migratory whales off the coast of the Pacific Northwest. Focus on your local flora and fauna, the plants and animals you already know. If you are interested in plants and have lots of daffodils and apple trees, then you could focus on these and record when they first flower and when they finish flowering in the spring. Or you might record the appearance of dandelions in your lawn. If you are interested in birds, you could record when each species of bird arrives in the spring. Or perhaps mushrooms or dragonflies are your passion. Even lichens on local gravestones may be responding to climate change. Whatever it is, start recording your observations in your journal.

Another advantage of keeping a journal is that it gets you out in nature. If you are reading this book, chances are you are already a member of the choir for this particular sermon. But even the most ardent outdoor enthusiasts among us can feel the tug of work, digital devices, and a busy schedule. A journal can be a way to make sure we make time for nature in our twenty-first-century lives. With journals in hand, weekends and vacations can become times when you join with family and friends for walks in parks and nature reserves, observing the changes in the seasons. An afternoon watching a heron rookery or climbing a nearby hill to make observations of leafing out and other aspects of nature will prove more memorable than spending that time in front of a computer screen or wandering around a shopping mall. We know life spent closer to nature, studying it and observing it, is a richer life than most of us pursue. A nature journal can help us act on that knowledge.

Joining with Others

Outings with environmental groups and natural history clubs also offer a good way to improve identification skills, especially of migratory songbirds such as warblers, which can be difficult to tell apart without practice. In the Boston area, we have clubs for the study and observation of birds, plants, insects, and mushrooms, and even more specialized clubs, such as the Massachusetts club for butterfly enthusiasts detailed in chapter 9. Such groups have frequent field trips during the most active seasons and are also a good way to meet people with common interests. Join the club if you want to find out where to find the butterflies and how to identify them; you can talk for hours with dozens of members who are extremely knowledgeable about butterflies. If there isn't a naturalist's club where you live, you could contact your local university's extension program or nature museum to get help in establishing one.

You can also join existing online networks of observers. Many such networks already exist, and more are forming all of the time. One of the best established is Journey North (http://www.learner.org/jnorth/), which tracks the migrations of monarch butterflies, ruby-throated hummingbirds, and other migratory species that extend their range northward in the spring. Another is eBird, a program run by the Cornell Lab of Ornithology. In this program, observers from across North America enter their observations of birds in the spring into an online database; these observations help researchers to track the changing abundance and distribution of birds over the years, and in turn eBird provides the observers with checklists of all of the birds that they have seen, where they have seen the birds, and when they have seen each species.

Two related efforts are the USA National Phenology Network, which gathers rigorous data on plants and animals to address scientific questions, and Project BudBurst, which focuses on plants and has a greater emphasis on public education. Both established in 2007, these two networks track the appearance of hundreds of different plant and animal species, which means you can probably find at least a few to track if you live in North America. The disadvantage of these new systems is that they are relatively young enterprises and will take some years before they are able to show patterns of change.

In these online systems, you need to register and provide information on your geographic location, the type of habitat where you are observing, and whether your location is sunny or shady and if it is wet or dry. Once past these first steps, you are asked to provide information on such plant characters as when leaves first appear, when flowers open, when leaves change color and fall, and when fruit ripens. For birds, observers provide information on the date of observation of a certain species: how many birds, what they eat, and when they mate and build their nests.

People who start nature journals to keep track of the seasons often comment on how much pleasure it gives them. Now, with enormous interest in the effects of climate change on the environment, such journals can also make an important scientific contribution. These journals increase in value with the addition of each new year. While long-term records are the most valuable, important new insights often can come from enthusiastic observers just starting out. And with all of the rapid changes and fluctuations in the climate happening now and predicted for future years, this a great time to start a journal, both as an individual observer and as a contributor to a global network of concerned citizens.

MY OWN JOURNAL

My own journals have been more varied, ranging from terse records of data to almost personal diaries with some science woven in. Here is the entry for my first day of fieldwork in Concord when I first started this project on climate change:

> April 10, 2003: I went out today on a crisp spring day with snow still on the ground. I visited Great Meadow (but could only walk around part of it due to flooding), Estabrook Woods, and the Old Rifle Range. The only species observed in flower were red maple, skunk cabbage, and hazelnut that had been flowering earlier. Hazelnut flowers had been killed by recent frost. I am still not sure where the best places will be to look for plants in this study.

A year later, I was starting to develop a better idea of what I needed for my research project:

> April 8, 2004: Today is the first day of the field season. I feel optimistic and excited. We know so much more than last year. And today is a beautiful spring day; cool, around 40 degrees, sunny, and totally clear.
>
> Punkatasset. In flower: alder, skunk cabbage.
>
> Sleepy Hollow Cemetery. Nothing in flower.

After eight more years, my journal had gotten more precise, as I knew better what type of data was required for my climate change study.

April 1, 2012: Concord, 10:30 a.m. Sunny, 45 degrees. By myself.

At Walden Pond. Leafing out for the first time: early lowbush blueberry, black cherry, highbush blueberry, shadbush.

In flower for the first time: marsh marigold, spicebush.

Minute Man National Historical Park. In flower for the first time: yellow violets, dandelion, periwinkle.

April 2, 2012: Last night, the temperature declined to 24 degrees, killing many flowers and young leaves.

Appendix: Species Mentioned

Wildflowers and Trees

COMMON NAME	SCIENTIFIC NAME
Arethusa orchid	*Arethusa bulbosa*
Asian bittersweet§	*Celastrus orbiculatus*
Aster	*Aster spp.*
Bayberry	*Myrica pennsylvanica*
Beech	*Fagus grandifolia*
Big-leaf magnolia	*Magnolia macrophylla*
Black birch	*Betula lenta*
Black cherry	*Prunus serotina*
Black swallowwort§	*Cynanchum louisease*
Bladder campion	*Silene cucubalus*
Bluet	*Houstonia caerulea*
Bulbous buttercup	*Ranunculus bulbosus*
Bur marigold	*Bidens cernua*
Buttercup	*Ranunculus spp.*
Canada lily	*Lilium canadense*
Canada mayflower	*Maianthemum canadense*
Cardinal flower	*Lobelia cardinalis*
Carolina fanwort§	*Cabomba caroliniana*
Catbriar	*Smilax rotundifolia*
Chokeberry	*Aronia arbutifolia*
Early lowbush blueberry	*Vaccinium angustifolium*
Elm	*Ulmus americana*
False hellebore	*Veratrum viride*
False rue-anemone	*Isopyrum biternatum*
Fire cherry	*Prunus pensylvanica*
Fringed gentian	*Gentianopsis crinita*
Fringed polygala	*Polygala paucifolia*

COMMON NAME	SCIENTIFIC NAME
Garden lupine	*Lupinus polyphyllus*
Garlic mustard§	*Alliaria petiolata*
Goldenrod	*Solidago spp.*
Goldenthread	*Coptis groenlandica*
Green-fringed orchid	*Habenaria lacera*
Highbush blueberry	*Vaccinium corymbosum*
Huckleberry	*Gaylussacia baccata*
Humped bladderwort	*Utricularia gibba*
Japanese knotweed§	*Polygonum cuspidatum*
Knotweed (native)	*Polygonum spp.*
Lady's slipper orchid	*Cypripedium acaule*
Large-flowered trillium	*Trillium grandiflorum*
Lily-of-the-valley	*Convallaria majalis*
Longflower tobacco	*Nicotiana longiflora*
Marsh mallow	*Althaea officinalis*
Marsh marigold	*Caltha palustris*
Meadow beauty	*Rhexia virginica*
Mountain mint	*Pycnanthemum spp.*
New England aster	*Aster novae-angliae*
Nodding lady's tresses orchid	*Spiranthes cernua*
Ovate-leaved violet	*Viola fibriatula*
Purple-flowered bladderwort	*Utricularia purpurea*
Purple-fringed orchis	*Habenaria psychodes*
Purple loosestrife§	*Lythrum salicaria*
Red maple	*Acer rubrum*
Rhodora	*Rhododendron canadense*
Rose	*Rosa spp.*
Rose pogonia	*Pogonia ophioglossoides*
Roseshell azalea	*Rhododendron prinophyllum*
Shadbush	*Amelanchier spp.*
Silky dogwood	*Cornus amomum*
Silver maple	*Acer saccharinum*
Slender lady's tresses orchid	*Spiranthes gracilis*

COMMON NAME	SCIENTIFIC NAME
Spicebush	*Lindera benzoin*
Stinking chamomile (mayweed)	*Anthemis cotula*
Swamp milkweed	*Asclepias incarnata*
Trailing arbutus	*Epigaea repens*
Velvetleaf	*Abutilon theophrasti*
Water chestnut[§]	*Trapa natans*
White oak	*Quercus alba*
Whorled pogonia	*Isotria verticillata*
Whorled wood aster	*Aster acuminatus*
Wild apple	*Malus domestica*
Willow herb	*Epilobium glandulosum*
Wood anemone	*Anemone quinquefolia*
Yellow birch	*Betula alleghaniensis*
Yellow root	*Xanthorhiza simplicissima*
Yellow wood sorrel	*Oxalis europaea*

[§] Non-native invasive species.

Birds Mentioned by Common Name

COMMON NAME	SCIENTIFIC NAME
American robin	*Turdus migratorius*
Baltimore oriole	*Icterus galbula*
Barn swallow	*Hirundo rustica*
Black-poll warbler	*Dendroica striata*
Blue jay	*Cyanocitta cristata*
Bobolink	*Dolichonyx oryzivorus*
Brown thrasher	*Toxostoma rufum*
Catbird	*Dumetella carolinensis*
Chipping sparrow	*Spizella passerina*
Eastern meadowlark	*Sturnella magna*
Eastern phoebe	*Sayornis phoebe*
Eastern towhee	*Pipilo erythrophthalmus*

COMMON NAME	SCIENTIFIC NAME
Eastern wood-peewee	*Contopus virens*
Field sparrow	*Spizella pusilla*
Fox sparrow	*Passerella iliaca*
Great blue heron	*Ardea herodias*
Great-crested flycatcher	*Myiarchus crinitus*
Hermit thrush	*Catharus guttatus*
House wren	*Troglodytes aedon*
Indigo bunting	*Passerina cyanea*
Ovenbird	*Seiurus aurocapilla*
Pine warbler	*Dendroica pinus*
Purple finch	*Carpodacus purpureus*
Red-eyed vireo	*Vireo olivaceous*
Red-winged blackbird	*Agelaius phoeniceus*
Ruby-throated hummingbird	*Archilocus colubris*
Savannah sparrow (Thoreau's "seringo")	*Passerculus sandwichensis*
Swamp sparrow	*Melospiza Georgiana*
Tree swallow	*Tachycineta bicolor*
Warbling vireo	*Vireo gilvus*
Wood duck	*Aix sponsa*
Wood thrush	*Hylocichla mustelina*
Yellow-rumped warbler	*Dendroica coronata*
Yellow warbler	*Dendroica petechia*

Insects Mentioned

COMMON NAME	SCIENTIFIC NAME
BUTTERFLIES	
Cabbage white	*Pieris rapae*
Eastern pine elfin	*Callophrys niphon*
Frosted elfin	*Callophrys irus*
Mourning cloak	*Nymphalis antiopa*
Pearl crescent	*Phyciodes tharos*
Peck's skipper	*Polites peckius*

COMMON NAME	SCIENTIFIC NAME
Red admiral	*Vanessa atalanta*
Spring azure	*Celastrina ladon*
MOSQUITOES	CULICIDAE
Black-tailed mosquito	*Culiseta melanura*
	Aedes spp.
	Culex spp.
OTHER INSECTS	
Caddis flies	Trichoptera
Crane flies	Tipulidae
Diving beetle	*Thermonectus spp.*
Dragonflies	Odonata
Giant water beetle	*Lethocerus americanus*
Gnats	Small flies in the order Diptera
Mayflies	Ephemeroptera
Non-biting Midges	*Chironomus spp.* and others
Stoneflies	Plecoptera
Water strider	Gerridae

Amphibians and Reptiles Mentioned

COMMON NAME	SCIENTIFIC NAME
SALAMANDERS	
Blue-spotted salamander	*Ambystoma laterale*
Red-backed salamander	*Plethodon cinereus*
Yellow-spotted salamander	*Ambystoma maculatum*
FROGS	
American toad	*Bufo americanus*
Bullfrog	*Rana catesbeiana*

COMMON NAME	SCIENTIFIC NAME
Spring peeper	*Pseudacris crucifer*
Wood frog	*Rana sylvatica*
TURTLES	
Blanding's turtle	*Emydoidea blandingii*

Acknowledgments

The work presented here was carried out by numerous Boston University students, most notably the graduate students Abe Miller-Rushing, Caroline Polgar, and Libby Ellwood, and undergraduates Caroline Imbres, Anna Ledneva, Dan Primack, Luba Zhaurova, Kiruba Dharaneeswaran, Sharda Mukunda, Paul Satzinger, and Michelle Talmadge, though others participated as well. Information on Concord field sites was generously provided by Mary Walker, Bryan Windmiller, Peter Alden, Ray Angelo, Susan Clark, and other Concord residents. Brad Dean and Phil Cafaro pointed out the Thoreau data sets that formed the basis for much of this work. Many other people mentioned in the text and in our publications worked on specific research projects, in particular Chuck Davis, Charlie Willis, Brad Ruhfel, Sharon Stichter, Ernest Williams, Trevor Lloyd-Evans, Kathleen Anderson, Nathan Phillips, Robert Kaufmann, and Colleen Hitchcock. Funding for this research was provided by the National Science Foundation and Boston University. The historical resources provided by libraries were critical to this project; special thanks are due to Leslie Perrin Wilson and Connie Manoli-Skocay of the Concord Free Public Library. Fellowships for writing came from the Guggenheim Foundation and the Arnold Arboretum of Harvard University. I appreciate the many colleagues, friends, and relatives who read and commented on chapters of the book, especially Elizabeth Platt, Phil Cafaro, Margaret Primack, Allan Chavkin, Nancy Chavkin, Don Lyman, Mark Primack, Bernd Heinrich, Rika Fuchino, Caitlin McDonough MacKenzie, Amanda Gallinat, Meg Boeni, and Hal Ober. Ann Downer-Hazell from Elefolio worked with me to edit the book into its present form. More details of the projects mentioned in the book and the original scientific papers can be found on our lab website (http://people.bu.edu/primack/). And finally I thank my wife, Margaret, and our children for their encouragement and support over the years.

Further Reading

Walden Warming is largely based on the work that we have been conducting in Concord and elsewhere in Massachusetts for the past twelve years. The scientific articles that we have written can be found on our lab website. The website is frequently updated and will have our most recent work as well:

http://people.bu.edu/primack/

We also provide regular updates on our lab blog, titled *Climate Change Research in Partnership with Thoreau*:

http://primacklab.blogspot.com/

Much of our work is summarized in the following review article:

Primack, R. B., and A. J. Miller-Rushing. "Uncovering, Collecting and Analyzing Records to Investigate the Ecological Impacts of Climate Change: A Template from Thoreau's Concord." *BioScience* 62 (2012): 170–81.

Articles about work are also presented on the BU Public Relations website:

http://www.bu.edu/news/category/richard-primack/

A five-minute video about our work is available from *BU Today*:

http://www.bu.edu/today/2011/watching-climate-change-from-the-ground-and-the-heavens/

Many of my ideas on the conservation of biological diversity are further developed in the following two books:

Primack, Richard B. *Essentials of Conservation Biology*. 5th ed. Sunderland, MA: Sinauer Associates, 2010.
Primack, Richard B. *A Primer of Conservation Biology*. 5th ed. Sunderland, MA: Sinauer Associates, 2012.

For those interested in going deeper into the literature, here are some readings on particular topics.

On Global Warming and Climate Change

Cafaro, Philip. "Beyond Business as Usual: Alternative Wedges to Avoid Catastrophic Climate Change and Create Sustainable Societies," In *The Ethics of Global Climate Change*, edited by Denis Arnold. Cambridge: Cambridge University Press, 2011.

Caldeira, K. "The Great Climate Experiment." *Scientific American* 307, no. 3 (September 2012): 78–83.

Flannery, Tim. *The Weather Makers: Our Changing Climate and What It Means for Life on Earth*. New ed. New York: Penguin, 2007.

McKibben, Bill. *Eaarth: Making a Life on a Tough New Planet*. New York: St. Martin's/Griffin, 2011.

Miller, P. "Weather Gone Wild." *National Geographic* 222, no. 3 (September 2012): 30–55.

Newman, J. A., et al. *Climate Change Biology*. Oxfordshire, UK: CAB International, 2011.

Weart, Spencer. *The Discovery of Global Warming*. Rev. and expanded ed. (New Histories of Science, Technology, and Medicine). Cambridge, MA: Harvard University Press, 2008.

By and about Thoreau

Johnson, L. *Thoreau's Complex Weave: The Writing of a Week on the Concord and Merrimack Rivers*. Charlottesville: University Press of Virginia, 1986.

Oehlschlaeger, F., and G. Hendrick, eds. *Toward the Making of Thoreau's Modern Reputation*. Urbana: University of Illinois Press, 1979.

Richardson, Robert D., Jr. *Henry Thoreau: A Life of the Mind*. Berkeley: University of California Press, 1988.

Thoreau, Henry David. *Walden: A Fully Annotated Edition*. Edited by Jeffrey S. Cramer. New Haven, CT: Yale University Press, 2004.

Thoreau, Henry David. *The Journal of Henry David Thoreau*. Edited by Bradford Torrey and Francis H. Allen. New York: Dover, 1962.

Thoreau, Henry David. *Walden*. Edited by Jeffrey S. Cramer. New Haven, CT: Yale University Press, 2006. (Page numbers in the text are from this excellent edition; many others are available.)

Thoreau, Henry David. *A Week on the Concord and Merrimack Rivers / Walden; Or, Life in the Woods / The Maine Woods / Cape Cod*. New York: Library of America, 1989.

Learning the Flora and Fauna

Alden, Peter. *National Audubon Society Field Guide to New England*. New York: Random House, 1998.

Burk, John S. *The Wildlife of New England: A Viewer's Guide*. Lebanon: University of New Hampshire Press, 2011.

Gracie, Carol. *Spring Wildflowers of the Northeast: A Natural History*. Princeton, NJ: Princeton University Press, 2012.

Haines, Arthur. *New England Wild Flower Society's* Flora Novae Angliae*: A Manual for the Identification of Native and Naturalized Higher Vascular Plants of New England*. New Haven, CT: Yale University Press, 2011.

Sibley, David Allen. *The Sibley Field Guide to Birds of Eastern North America*. New York: Knopf Doubleday, 2003.

The Peterson Field Guides (Houghton Mifflin) and National Audubon Society Field Guides are both good and are available in regional editions for birds and wild plants, as well as national editions for insects, amphibians and reptiles, mammals, and fishes, in print as well as apps for smartphones and tablet devices.

On Observing Nature

Leslie, Clare Walker. *Keeping a Nature Journal: Discover a Whole New Way of Seeing the World Around You*. North Adams, MA: Storey, 2003.

Williams, Ernest Herbert. *The Nature Handbook: A Guide to Observing the Great Outdoors*. New York: Oxford University Press, 2005.

Climate Change and Disease

Epstein, Paul R., and Dan Ferber. *Changing Planet, Changing Health: How the Climate Crisis Threatens Our Health and What We Can Do about It*. Berkeley: University of California Press, 2011.

The Physiology of Running

Heinrich, Bernd. *Why We Run: A Natural History*. New York: Harper Perennial, 2002.

Living Green and Lean

Berners-Lee, Mike. *How Bad Are Bananas?: The Carbon Footprint of Everything*. Vancouver, BC: Greystone, 2011.

Bongiorno, Lori. *Green, Greener, Greenest: A Practical Guide to Making Eco-Smart Choices Part of Your Life*. New York: Perigree, 2008.

Brown, Azby. *Just Enough: Lessons in Living Green from Traditional Japan*. New York: Kodansha, 2010.

Graaf, John de, David Wann, and Thomas H. Naylor. *Affluenza: The All-Consuming Epidemic*. San Francisco: Berrett-Koehler, 2005.

Pollan, Michael. *Food Rules: An Eater's Manual*. New York: Penguin, 2010.

Pollan, Michael. *The Omnivore's Dilemma: A Natural History of Four Meals*. New York: Penguin, 2006.

Union of Concerned Scientists. *Cooler Smarter: Practical Steps for Low-Carbon Living*. Washington DC: Island Press, 2012.

Index

Boston Marathon: bombing at, 209; effect of temperature on, 199, 201–9; effect of wind on, 206; history of, 204; picture of, 197; Primack's running of, 199–201, 208

Boston University: museum collections of, 146; Primack's career at, 4, 84; students from, 32, 100, 126, 137

Brewster, William, 100, 102–7, 110

budburst, flowering. *See* flowering times (dates); leaf out

buttercups, 30, 68, 70

butterflies: changing flight times of, 215; migration of, 133, 178, 232; as part of an ecosystem, 134; picture of, 76; as pollinators, 68; records of observations of, 124–26, 137–50; Thoreau quotation about, 132. *See also* caterpillars; insects; Massachusetts Butterfly Club; monarchs

cabbage white butterfly, 139–40

Cafaro, Phil, 4, 220, 222, 241

Cambridge, MA, 33, 100, 102, 104, 199–200

Cape Cod, 103, 117, 171, 213

carbon dioxide: atmospheric concentration of, 12, 79, 214, 218, 225; mosquito detection of, 167, 172; oceanic concentrations of, 216; plant uptake of, 1, 55, 79, 82, 215; reduction in, 220–21; release of, 1, 7, 219. *See also* fossil fuels; greenhouse gases

Carbon Mitigation Initiative, 219–20

Carolina fanwort, 64

caterpillars, 53–54, 95, 133, 139, 141, 143–45. *See also* butterflies

Charles River, 8, 101–2, 213

cherry trees, 34, 72, 130, 160, 207, 234

chestnut trees, 61

Christmas Bird Count, 96, 104. *See also* bird-watching

citizen science, 100, 135, 150, 229–33

civil disobedience, 29. See also *On Civil Disobedience* (Thoreau)

Clark, Susan, 61, 241

clovers, 47, 49, 145

Conantum (neighborhood), 62

Concord Free Public Library, 26, 48, 56, 151

Concord landfill, 86

Concord River, 30, 41–42, 82, 96

conservation: of animals, 2, 181; areas, 31, 82, 89, 96, 191; of energy resources, 209; field of, 2, 118; management, 43, 89, 194; organizations, 100, 156, 181; practitioners of, 5, 123, 128

Corey, Rosita, 104–7, 110, 123

Cornell Lab of Ornithology, 97, 232

Cramer, Jeff, 5, 101

Davis, Chris, 43

Davis, Chuck, 66–68, 73, 79–80, 241

DDT. *See* pesticides

Dean, Brad, 5

deer: as game animals, 7, 28, 60–61; as grazers, 40, 49, 66, 74–75

DeMaria, Alfred, 169

ducks, 11, 96, 97, 124, 125–26, 212

Duke University, 2

Eastern equine encephalitis, 170–77

Eaton, Richard, 62–64, 70–72, 85

ebird. *See* Cornell Lab of Ornithology

Edwards, Scott, 96

elfin butterflies, 131, 144–50

Ellwood, Libby, 101–12, 122–23, 126, 169, 241

Emerson, Ralph Waldo, 129, 135, 198

endangered species: amphibians, 181, 182–83, 193, 195; orchids, 62; protecting populations of, 88, 191; threats to from climate change, 89, 91, 216–17

Ernst Mayer Library, 101–2

Estabrook Woods, 40, 87, 233

extinction, general, 61, 70, 81, 141, 181, 217

human health: effect of decreased material consumption on, 221, 222, 225;
effect of diet on, 215, 220; effect of
temperature on, 217; vector-borne
disease, 162, 165–66, 169–71, 175–78.
See also Eastern equine encephalitis;
West Nile virus
hummingbirds: general, 1, 68, 117, 231;
ruby-throated, 124, 125, 129, 133, 232
hunting, 28, 61, 65, 74, 158

ice: fishing, 201; harvesting, 12, 24, 28;
ice-out, 5, 10–13, 17–18, 29, 125, 156,
216; skating, 8–9, 189, 198, 201; thickness of, 7–10
indigo bunting, 107
Industrial Revolution, 7, 12, 158
insects: aquatic (not mosquitoes), 11,
153–60, 162, 184–85, 177, 189; citizen
science observations of, 230, 232; as
controls of invasive plants, 79; effect
of climate change on, 30, 90, 108,
128–29, 162; life cycle of, 133–34;
as part of a food web, 53–55, 69,
97–99, 111, 121, 127, 150, 187, 193;
predictions for the future of, 212, 216;
search for data on, 137–42, 146, 168;
thermoregulation of, 202; Thoreau's
observations of, 95, 152; threats to,
177–78, 194. *See also* bees; butterflies;
caterpillars; mosquitoes
Intergovernmental Panel on Climate
Change, 16, 218
International Union for Conservation of
Nature, 218
invasive species: aquatic plants, 64;
flowering-time flexibility, 79–81;
future problems with, 212; introduction of, 78–79; relationship to
overwintering birds, 174; threatening rare species, 40, 41, 66; threats
associated with assisted colonization, 91

jack-in-the-pulpit, 21
Janzen, Dan, 2
Japan, 74, 222–24
jays, 99, 116, 118
Journey North (website), 232

Korea, 121

land-use change, 1, 65, 74. *See also* agriculture; logging
leaf out: data from Primack's journal on,
234; effect of temperature on, 54, 70,
88, 229; effect on plant growth, 55;
flexibility of, 69–70, 80; future predictions of, 212; planning field work,
4; in relation to other organisms, 54,
98, 140; suggestions for observing,
231; Thoreau's observations of, 5, 29,
207
lilacs, 86, 230
lilies: Canada, 40, 43, 69; general, 60, 67;
lily-of-the-valley, 87
Lloyd-Evans, Trevor, 100, 115–16, 122–24,
126, 241
logging, 1, 28, 55, 61, 65, 119
lynx, 60, 64

magnolias, 88, 149
Malaysia, 1–3, 159
Manomet Center for Conservation Sci-
ence, 100, 113, 115–23
maples: general, 41, 152, 160; red, 21, 42,
53, 129, 171, 189, 233; silver, 42; sugar,
212
marathons, effect of climate change on.
See Boston Marathon
marsh marigold, 33, 49, 52, 234
Martha's Vineyard, 148, 160
Massachusetts Butterfly Atlas, 143–44
Massachusetts Butterfly Club, 138–39,
142–44, 145, 147–50, 232
Massachusetts Division of Fisheries and
Wildlife, 191

28, 44, 63, 66; picture of, 14, 93, 210; plants and animals of, 40, 156–57, 185–86; predicted future of, 212, 214–15; records from, 55; research and observations at, 16, 25, 29–31, 34, 49, 227; temperature of, 155–56; Thoreau's time at, 5, 7, 24, 85, 226; visitors to, 158; water level of, 18–19, 186

Walker, Mary, 61, 241

warblers: decline in population of, 89; habits of, 127; historical mentions of, 103, 132; identification of, 105, 116, 231; spring arrival of, 107, 111, 120

water chestnut, 64

water lilies, 64

West Concord, 34, 64, 86, 104

West Nile virus, 165, 169, 175, 178

White, Gilbert, 229

Williams, Ernest, 131, 147, 241

Willis, Charlie, 66–68, 75, 79, 241

Windmiller, Bryan, 61, 182–83, 241

wolves, 60–61, 64, 74, 203

wood sorrel, 47, 83

wood thrush, 107

wrens, 125

Wyman's Meadow, 185–86